UAV AIRCRAFT LOGBOOK

PRO

AIRCRAFT:

Logbook number _____

Entries from _____

Through _____

AIRCRAFT OWNER / OPERATOR

Owner

Name

Address

Telephone Number

Change of Address / Telephone Number

Operator

Name

Address

Telephone Number

Change of Address / Telephone Number

CONTENTS

Notes on the use of this book:

Use Part 1 to record aircraft and equipment characteristics.

Use Part 2 to record daily flights and any mechanical defects or incidents that occurred before or during flight and to record any repairs made.

Use Part 3 to record such actions as: planned maintenance, equipment upgrades, firmware updates and sensors/equipment calibrations (other than compass calibrations, which should be done as part of the pre-flight checklist).

PART 1: AIRCRAFT, POWER SOURCES AND EQUIPMENT

AIRCRAFT DATA

Aircraft Identity		Aircraft Location	
Make		Address	
Model			
Year of Manufacture			
Date Acquired			
Registration Number		Change of Location	
Registration Authority			
Date of Registration		Address	
Place of Registration			
Type (e.g. Fixed-wing, Rotary, LTA):			
Power (e.g. Batteries, Liquid Fuel, Unpowered):			

9

Aircraft Insurance

Insurance Provider

Policy Type / Name

Policy Number and Date Acquired

Remarks (e.g. amount insured, types of damage insured, etc)

Names of the pilots who are authorized to fly the insured aircraft

AIRCRAFT DATA, CONTINUED

List of Special Approvals Associated With This Registration/Aircraft (e.g. FAA night flying waiver)

1
2
3
4
5
6
7
8
9
10
11
12

AIRCRAFT DATA, CONTINUED

List of Special Approvals Associated With This Registration/Aircraft (e.g. FAA night flying waiver)

13	
14	
15	
16	
17	
18	
19	
20	
21	
22	
23	
24	

CONTROLLERS DATA

(Marking "Out," with checkbox, designates a device no longer in use)

Controller Identity

Designation (e.g. name or number)

Make

Model

Year of Manufacture

Date Acquired

☐ Out

Controller Identity

Designation (e.g. name or number)

Make

Model

Year of Manufacture

Date Acquired

☐ Out

Controller Identity

Designation (e.g. name or number)

Make

Model

Year of Manufacture

Date Acquired

☐ Out

Controller Identity

Designation (e.g. name or number)

Make

Model

Year of Manufacture

Date Acquired

☐ Out

CONTROLLERS DATA, CONTINUED

(Marking "Out," with checkbox, designates a device no longer in use)

Controller Identity

Out ☐

Designation (e.g. name or number)

Make

Model

Year of Manufacture

Date Acquired

Controller Identity

Out ☐

Designation (e.g. name or number)

Make

Model

Year of Manufacture

Date Acquired

Controller Identity

Out ☐

Designation (e.g. name or number)

Make

Model

Year of Manufacture

Date Acquired

Controller Identity

Out ☐

Designation (e.g. name or number)

Make

Model

Year of Manufacture

Date Acquired

Aircraft Batteries Data

(24 batteries: Marking "Out," with checkbox, designates a battery no longer in use)

	Out:	Out:	Out:	Out:	Out:	Out:	Out:	Out:
Battery Designation								
Make								
Type or Model								
Date Acquired								
mAh / Vmax								

	Out:	Out:	Out:	Out:	Out:	Out:	Out:	Out:
Battery Designation								
Make								
Type or Model								
Date Acquired								
mAh / Vmax								

	Out:	Out:	Out:	Out:	Out:	Out:	Out:	Out:
Battery Designation								
Make								
Type or Model								
Date Acquired								
mAh / Vmax								

CONTROLLER BATTERIES, LIQUID FUEL AND BUOYANCY GAS DATA

(16 Batteries: Marking "Out," with checkbox, designates a battery no longer in use)

Batteries for (controller designation): _____

	Out: ☐	Out: ☐	Out: ☐	Out: ☐	Out: ☐	Out: ☐	Out: ☐	Out: ☐
Battery Designation								
Make								
Type or Model								
Date Acquired								
mAh / Vmax								

Batteries for (controller designation): _____

	Out: ☐	Out: ☐	Out: ☐	Out: ☐	Out: ☐	Out: ☐	Out: ☐	Out: ☐
Battery Designation								
Make								
Type or Model								
Date Acquired								
mAh / Vmax								

Aircraft Fuel (liquid-fueled UAV)

Type	
Source	
Full Tank Quantity	

Buoyancy Gas (LTA UAV)

Type	
Source	
Full Aircraft Quantity	

EQUIPMENT

(Marking "Out," with checkbox, designates a device no longer in use)

Make / Model / Identification Number / Purpose of Equipment Recorded in This Book (e.g. cameras, lenses, sensors)	Date Acquired	
1		Out: ☐
2		Out: ☐
3		Out: ☐
4		Out: ☐
5		Out: ☐
6		Out: ☐
7		Out: ☐
8		Out: ☐
9		Out: ☐
10		Out: ☐
11		Out: ☐
12		Out: ☐

EQUIPMENT, CONTINUED

(Marking "Out," with checkbox, designates a device no longer in use)

	Make / Model / Identification Number / Purpose of Equipment Recorded in This Book (e.g. cameras, lenses, sensors)	Date Acquired
13		☐ Out:
14		☐ Out:
15		☐ Out:
16		☐ Out:
17		☐ Out:
18		☐ Out:
19		☐ Out:
20		☐ Out:
21		☐ Out:
22		☐ Out:
23		☐ Out:
24		☐ Out:

MASTER PRE-FLIGHT AIRCRAFT INSPECTION CHECKLIST

List all pre-flight check items in accordance with the aircraft manufacturer's recommendations.

1   First flight of day only: ☐

2   First flight of day only: ☐

3   First flight of day only: ☐

4   First flight of day only: ☐

5   First flight of day only: ☐

6   First flight of day only: ☐

7   First flight of day only: ☐

8   First flight of day only: ☐

9   First flight of day only: ☐

10   First flight of day only: ☐

11   First flight of day only: ☐

12   First flight of day only: ☐

Master Pre-flight Aircraft Inspection Checklist, continued

List all pre-flight check items in accordance with the aircraft manufacturer's recommendations.

13	First flight of day only:
14	First flight of day only:
15	First flight of day only:
16	First flight of day only:
17	First flight of day only:
18	First flight of day only:
19	First flight of day only:
20	First flight of day only:
21	First flight of day only:
22	First flight of day only:
23	First flight of day only:
24	First flight of day only:

NOTES CONCERNING AIRCRAFT, POWER SOURCES AND EQUIPMENT

PART 2: AIRCRAFT DAILY LOG

Instructions for Recording Information in Part 2: Aircraft Daily Log

(an example is provided on page 23)

Data that can be entered in each Aircraft Daily Log page (the even-numbered pages in this book):

(as presented across the top of each page, from left to right)

Date   Enter the date of the flight(s) in the space provided in the upper right corner of the page. Each page records the flights of a single calendar day only.

Flt. No.   Enter the number of each flight (e.g. 1, 2, 3, or DF1, DF2, DF3, in accordance with individual or company requirements). This number is the reference used when writing a corresponding remark in the "Discrepancies / Incidents" section of the page.

P-f. I. / C. Cal.
Remark, using a crew-member's initials or a "y" for Yes, that a pre-flight inspection was accomplished. Remark also whether a compass calibration was performed or not.

Crew Names
Record the name of the pilot who operated the aircraft and that of any other crew member involved (e.g. the observer or camera operator).

Flight Location
Enter the take-off location and the landing location. Note that in the example on page 21 these are the same place.

Fuel / Battery
Enter amount of fuel uplifted and burned, or battery charge and discharge values (in volts or % charged), to include the battery designation recorded in the Aircraft Batteries Data pages (in the example on page 21 the battery designations are "A" and "B"). Record the amount (e.g. volts or fuel) used for each flight. For an aircraft that uses two batteries during flight, record the values of both in the same box (you will have to write small).

Flight Data
Record the takeoff and landing times (specify whether using Local or UTC time). Record also the duration of each flight, in hours and minutes (or tenths), the number of landings made per flight (e.g. if several touchdowns were accomplished without turning off the engines then a corresponding number of landings would be recorded for a given flight), the distance flown (specify units used) and the altitude (MSL) or height (AGL) achieved (specify units used).

Instructions for Recording Information in Part 2: Aircraft Daily Log

(an example is provided on page 23)

Data that can be entered in each Aircraft Daily Log page (the even-numbered pages in this book):

(as presented across the <u>middle</u> of each page, from left to right)

After inspecting the aircraft and the maintenance entries in this logbook the pilot can sign in the space provided if he/she accepts the vehicle as being safe to fly.

Batteries

Cycles this page: Count the cycles (1 discharge followed by a recharge before the next use) per battery from the columns above and record the totals per battery here.

Cycles brought forward: Enter the **"Total Cycles to date"** data from the **Batteries** section of the *previous page* in the space provided.

Total Cycles to date: Sum the "Cycles this page" and the "Cycles brought forward" entries and enter these sums into the space provided.

Flight data

Totals this page: Sum the flight data entries from the columns above and enter these sums in the spaces provided.

Totals brought forward: Enter the **"Totals to date"** data from the **Flight Data** section of the *previous page* in the spaces provided.

Totals to date: Sum the "Totals this page" and the "Totals brought forward" entries and enter these sums into the spaces provided.

INSTRUCTIONS FOR RECORDING INFORMATION IN PART 2: AIRCRAFT DAILY LOG

(an example is provided on page 23)

Data that can be entered in each AIRCRAFT DAILY LOG page (the even-numbered pages in this book):

(as presented across the <u>bottom</u> of each page, from left to right, making a record of discrepancies and incidents encountered during the flight or pre-flight)

Flt. No.   Enter the number of the flight (from the numbered flight above) during which a discrepancy or incident occurred.

Discrepancies / Incidents
Enter a complete description of any mechanical defect or incident that occurred during the numbered flights (or pre-flight).

Enter also a description of any repair or fix accomplished to resolve the problems. Comment about any test flight requirement.

(Use **Part 3: Additional Maintenance Actions Log** to record planned maintenance, equipment upgrades, data updates, etc.)

Approval R-T-F status
Sign to attest to the fact that the appropriate authority (e.g. the owner, pilot or mechanic) has determined that the aircraft is repaired and is ready to fly.

Data that can be entered in each AIRCRAFT DAILY LOG, CONTINUED page (the odd-numbered pages in this book):

Write relevant operations information about each of the up to five flights recorded on the associated flight data page above. These observations/comments can be about any issue that the writer deems useful to record for a given flight. If an accident or incident occurred, use this space to record pertinent details. Record notes concerning technical issues associated with this aircraft and its equipment.

AIRCRAFT DAILY LOG

DATE: 9 August 2018

Flt. No.	P-f. I. / C. Cal.	Crew Names		Location		Fuel / Battery			Flight Data					
		Pilot	Other	From	To	Takeoff V	Landing V	Used V	T/o Time Z	Ld. Time Z	Duration	Ld.	Distance km	Ht. ft
46	Y / Y	J. Smith	T. Jones		Diamond Head, Oahu	A: 17.1	14.6	2.5	11:00	11:15	0:15	1	0.80	200
47	Y / Y	J. Smith	T. Jones		Diamond Head, Oahu	B: 16.9	15.0	1.9	11:50	12:00	0:10	1	0.65	100

(Abbreviations: **P-f. I. / C. Cal.** = Pre-flight Inspection / Compass Calibration; **T/o** = Takeoff; **Ld.**= Landing(s); **Ht.** = Height)

			Flight Data		
Totals this page:			0:25	2	1.45
Totals brought forward:			14:10	45	32.20
Totals to date:			14:35	47	33.65

Batteries

Cycles this page:	A 1	B 1	
Cycles brought forward:	A 13	B 11 C 12 D 09	
Cycles to date:	A 14	B 12 C 12 D 09	

Pilot's acceptance prior to today's flight(s):

John Smith

Approval R-T-F status

Signature / Date

T. Wilson / 9 Aug 18

Discrepancies / Incidents

Flt. No.	Description of Technical Issue	Corrective Action Performed (if required)
46	Pilot discovered that a propeller blade had a 2 x 1 mm piece missing.	Prop removed and replaced before flight. Test during next flight.

EXAMPLE PAGE: The pilot and an additional crew member logged 2 flights of a battery-powered UAV on 9 August 2018. They performed pre-flight inspections, including compass calibrations, and indicated these (V = Yes) in the **P-f. I. / C. Cal.** column. A pre-flight mechanical problem was recorded, fixed and signed-off by the mechanic in the lower part of the form, referencing the affected flight. The pilot signed as accepting the aircraft as airworthy after the fix (in the middle left section of the page).

The pilot recorded the pre- and post-flight battery voltages, taking care to distinguish which battery was used each flight (designated A and B here), and the total volts consumed. The two battery cycles were then recorded in the **Batteries** section in the middle of the page, totals were brought forward from the previous page and the total **Cycles to Date** were calculated, per battery. The pilot chose to record local time (as opposed to UTC), battery power in volts, distances in kilometers and maximum heights in feet (labelled at the top). Flight data were totaled and entered in the middle-right section of the page.

AIRCRAFT DAILY LOG

DATE:

Flt. No.	P-f. I. / C. Cal.	Crew Names		Location		Fuel / Battery			T/o Time	Ld Time	Flight Data			
		Pilot	Other	From	To	Takeoff	Landing	Used			Duration	Ld	Distance	Ht.

Pilot's acceptance prior to today's flight(s): _____

Batteries — Cycles this page: _____ Cycles brought forward: _____ **Cycles to date:** _____

Flight Data — Totals this page: _____ Totals brought forward: _____ **Totals to date:** _____

Discrepancies / Incidents

Flt. No.	Description of Technical Issue	Corrective Action Performed (if required)	Approval R-T-F status — Signature / Date

Aircraft Daily Log, continued

Flt. No.	Notes (e.g. purpose of flight, weather, sensors used, camera settings, results, lessons learned)

AIRCRAFT DAILY LOG

DATE:

Flt. No.	P-f. I. / C. Cal.	Crew Names		Location		Fuel / Battery			Flight Data					
		Pilot	Other	From	To	Takeoff	Landing	Used	T/o Time	Ld Time	Duration	Ld	Distance	Ht.

Pilot's acceptance prior to today's flight(s):

Batteries

Cycles this page:

Cycles brought forward:

Cycles to date:

Flight Data

Totals this page:

Totals brought forward:

Totals to date:

Discrepancies / Incidents

Flt. No.	Description of Technical Issue	Corrective Action Performed (if required)	Approval R-T-F status
			Signature / Date

Aircraft Daily Log, continued

Flt. No.	Notes (e.g. purpose of flight, weather, sensors used, camera settings, results, lessons learned)

AIRCRAFT DAILY LOG

DATE: _____

Flt. No.	P-f. I. / C. Cal.	Crew Names		Location		Fuel / Battery			Flight Data						
		Pilot	Other	From	To	Takeoff	Landing	Used	T/o Time	Ld Time	Duration	Ld	Distance	Ht.

Pilot's acceptance prior to today's flight(s): _____

Batteries

Cycles this page: _____
Cycles brought forward: _____
Cycles to date: _____

Flight Data

Totals this page: _____
Totals brought forward: _____
Totals to date: _____

Discrepancies / Incidents

Flt. No.	Description of Technical Issue	Corrective Action Performed (if required)	Approval R-T-F status
			Signature / Date

Aircraft Daily Log, continued

Flt. No.	Notes (e.g. purpose of flight, weather, sensors used, camera settings, results, lessons learned)

AIRCRAFT DAILY LOG

DATE:

Flt. No.	P-f. l. / C. Cal.	Crew Names		Location		Fuel / Battery			T/o Time	Ld Time	Flight Data				Approval R-T-F status
		Pilot	Other	From	To	Takeoff	Landing	Used			Duration	Ld	Distance	Ht.	Signature / Date

Pilot's acceptance prior to today's flight(s):

Batteries | Flight Data

Cycles this page:	Totals this page:
Cycles brought forward:	Totals brought forward:
Cycles to date:	**Totals to date:**

Discrepancies / Incidents

Flt. No.	Description of Technical Issue	Corrective Action Performed (if required)

Flt. No.	Notes (e.g. purpose of flight, weather, sensors used, camera settings, results, lessons learned)

AIRCRAFT DAILY LOG

DATE:

| Flt. No. | P-f.I. / C. Cal. | Crew Names | | Location | | Fuel / Battery | | | Flight Data | | | | | | |
		Pilot	Other	From	To	Takeoff	Landing	Used	T/o Time	Ld Time	Duration	Ld	Distance	Ht.

Pilot's acceptance prior to today's flight(s):

Batteries

Cycles this page:

Cycles brought forward:

Cycles to date:

Flight Data

Totals this page:

Totals brought forward:

Totals to date:

Discrepancies / Incidents

| Flt. No. | Description of Technical Issue | Corrective Action Performed (if required) | Approval R-T-F status |
			Signature / Date

AIRCRAFT DAILY LOG, CONTINUED

Flt. No.	Notes (e.g. purpose of flight, weather, sensors used, camera settings, results, lessons learned)

AIRCRAFT DAILY LOG

DATE:

Flt. No.	P-f.l. / C. Cal.	Crew Names		Location		Fuel / Battery			Flight Data						Approval R-T-F status
		Pilot	Other	From	To	Takeoff	Landing	Used	T/o Time	Ld Time	Duration	Ld	Distance	Ht.	Signature / Date

Pilot's acceptance prior to today's flight(s):

Batteries

Cycles this page:
Cycles brought forward:
Cycles to date:

Totals this page:
Totals brought forward:
Totals to date:

Flight Data

Discrepancies / Incidents

Flt. No.	Description of Technical Issue	Corrective Action Performed (if required)

Aircraft Daily Log, continued

Flt. No.	Notes (e.g. purpose of flight, weather, sensors used, camera settings, results, lessons learned)

AIRCRAFT DAILY LOG

DATE: _____

Flt. No.	P-f. I. / C. Cal.	Crew Names		Location		Fuel / Battery			Flight Data						
		Pilot	Other	From	To	Takeoff	Landing	Used	T/o Time	Ld Time	Duration	Ld	Distance	Ht.

Pilot's acceptance prior to today's flight(s): _____

Batteries

Cycles this page: _____
Cycles brought forward: _____
Cycles to date: _____

Flight Data

Totals this page: _____
Totals brought forward: _____
Totals to date: _____

Discrepancies / Incidents

Flt. No.	Description of Technical Issue	Corrective Action Performed (if required)	Approval R-T-F status
			Signature / Date

Aircraft Daily Log, continued

Flt. No.	Notes (e.g. purpose of flight, weather, sensors used, camera settings, results, lessons learned)

AIRCRAFT DAILY LOG   DATE:

Flt. No.	P-f. l. / C. Cal.	Crew Names		Location		Fuel / Battery			Flight Data					
		Pilot	Other	From	To	Takeoff	Landing	Used	T/o Time	Ld Time	Duration	Ld	Distance	Ht.

Pilot's acceptance prior to today's flight(s):

Batteries — Flight Data

Cycles this page: ____ Totals this page:
Cycles brought forward: ____ Totals brought forward:
Cycles to date: ____ **Totals to date:**

Discrepancies / Incidents

Flt. No.	Description of Technical Issue	Corrective Action Performed (if required)	Approval R-T-F status — Signature / Date

Aircraft Daily Log, continued

Flt. No.	Notes (e.g. purpose of flight, weather, sensors used, camera settings, results, lessons learned)

AIRCRAFT DAILY LOG

DATE:

Flt. No.	P-f.l. / C. Cal.	Crew Names		Location		Fuel / Battery			Flight Data					
		Pilot	Other	From	To	Takeoff	Landing	Used	T/o Time	Ld Time	Duration	Ld	Distance	Ht.

Pilot's acceptance prior to today's flight(s):

Batteries	Cycles this page:				Flight Data	Totals this page:					
	Cycles brought forward:					Totals brought forward:					
	Cycles to date:					**Totals to date:**					

Discrepancies / Incidents

Flt. No.	Description of Technical Issue	Corrective Action Performed (if required)	Approval R-T-F status
			Signature / Date

AIRCRAFT DAILY LOG, CONTINUED

Flt. No.	Notes (e.g. purpose of flight, weather, sensors used, camera settings, results, lessons learned)

AIRCRAFT DAILY LOG

DATE:

Flt. No.	P-f. I. / C. Cal.	Crew Names		Location		Fuel / Battery					Flight Data				
		Pilot	Other	From	To	Takeoff	Landing	Used	T/o Time	Ld Time	Duration	Ld	Distance	Ht.

Pilot's acceptance prior to today's flight(s): _____

Batteries

Cycles this page:
Cycles brought forward:
Cycles to date:

Flight Data

Totals this page:
Totals brought forward:
Totals to date:

Discrepancies / Incidents

Flt. No.	Description of Technical Issue	Corrective Action Performed (if required)	Approval R-T-F status	
			Signature / Date	

Aircraft Daily Log, continued

Flt. No.	Notes (e.g. purpose of flight, weather, sensors used, camera settings, results, lessons learned)

AIRCRAFT DAILY LOG

DATE: _____

Flt. No.	P-f. I. / C. Cal.	Crew Names		Location		Fuel / Battery			Flight Data					
		Pilot	Other	From	To	Takeoff	Landing	Used	T/o Time	Ld Time	Duration	Ld	Distance	Ht.

Pilot's acceptance prior to today's flight(s): _____

Batteries
Cycles this page: _____
Cycles brought forward: _____
Cycles to date: _____

Flight Data
Totals this page: _____
Totals brought forward: _____
Totals to date: _____

Discrepancies / Incidents

Flt. No.	Description of Technical Issue	Corrective Action Performed (if required)	Approval R-T-F status	Signature / Date

AIRCRAFT DAILY LOG, CONTINUED

Flt. No.	Notes (e.g. purpose of flight, weather, sensors used, camera settings, results, lessons learned)

AIRCRAFT DAILY LOG

DATE:

Flt. No.	P-f.I. / C. Cal.	Crew Names		Location		Fuel / Battery			Flight Data					
		Pilot	Other	From	To	Takeoff	Landing	Used	T/o Time	Ld Time	Duration	Ld	Distance	Ht.

Pilot's acceptance prior to today's flight(s):

		Batteries	Flight Data	
Cycles this page:		Totals this page:	
Cycles brought forward:		Totals brought forward:	
Cycles to date:		**Totals to date:**	

Discrepancies / Incidents

Flt. No.	Description of Technical Issue	Corrective Action Performed (if required)	Approval R-T-F status — Signature / Date

46

AIRCRAFT DAILY LOG, CONTINUED

Flt. No.	Notes (e.g. purpose of flight, weather, sensors used, camera settings, results, lessons learned)

AIRCRAFT DAILY LOG

DATE:

Flt. No.	P-f. I. / C. Cal.	Crew Names		Location		Fuel / Battery			Flight Data					
		Pilot	Other	From	To	Takeoff	Landing	Used	T/o Time	Ld Time	Duration	Ld	Distance	Ht.

Flight Data

Totals this page:
Totals brought forward:
Totals to date:

Batteries

Cycles this page:
Cycles brought forward:
Cycles to date:

Pilot's acceptance prior to today's flight(s):

Discrepancies / Incidents

Flt. No.	Description of Technical Issue	Corrective Action Performed (if required)	Approval R-T-F status
			Signature / Date

Flt. No.	Notes (e.g. purpose of flight, weather, sensors used, camera settings, results, lessons learned)

AIRCRAFT DAILY LOG

DATE:

Flt. No.	P-f.I. / C. Cal.	Crew Names		Location		Fuel / Battery			T/o Time	Ld Time	Flight Data			
		Pilot	Other	From	To	Takeoff	Landing	Used			Duration	Ld	Distance	Ht.

Pilot's acceptance prior to today's flight(s):

Batteries

Cycles this page:
Cycles brought forward:
Cycles to date:

Flight Data

Totals this page:
Totals brought forward:
Totals to date:

Discrepancies / Incidents

Flt. No.	Description of Technical Issue	Corrective Action Performed (if required)	Approval R-T-F status
			Signature / Date

Flt. No.	Notes (e.g. purpose of flight, weather, sensors used, camera settings, results, lessons learned)

AIRCRAFT DAILY LOG

DATE:

52

Flt. No.	P-f. I. / C. Cal.	Crew Names		Location		Fuel / Battery			T/o Time	Ld Time	Flight Data			
		Pilot	Other	From	To	Takeoff	Landing	Used			Duration	Ld	Distance	Ht.

Flight Data

Totals this page:
Totals brought forward:
Totals to date:

Batteries

Cycles this page:
Cycles brought forward:
Cycles to date:

Pilot's acceptance prior to today's flight(s): _____

Discrepancies / Incidents

Flt. No.	Description of Technical Issue	Corrective Action Performed (if required)	Approval R-T-F status — Signature / Date

AIRCRAFT DAILY LOG, CONTINUED

Flt. No.	Notes (e.g. purpose of flight, weather, sensors used, camera settings, results, lessons learned)

AIRCRAFT DAILY LOG

DATE: _____

Flt. No.	P-f. I. / C. Cal.	Crew Names		Location		Fuel / Battery			Flight Data					
		Pilot	Other	From	To	Takeoff	Landing	Used	T/o Time	Ld Time	Duration	Ld	Distance	Ht.

Pilot's acceptance prior to today's flight(s):

Batteries

Cycles this page: _____
Cycles brought forward: _____
Cycles to date: _____

Flight Data

Totals this page: _____
Totals brought forward: _____
Totals to date: _____

Discrepancies / Incidents

Flt. No.	Description of Technical Issue	Corrective Action Performed (if required)	Approval R-T-F status — Signature / Date

AIRCRAFT DAILY LOG, CONTINUED

Flt. No.	Notes (e.g. purpose of flight, weather, sensors used, camera settings, results, lessons learned)

AIRCRAFT DAILY LOG

DATE:

Flt. No.	P-f.I. / C. Cal.	Crew Names		Location		Fuel / Battery			Flight Data					
		Pilot	Other	From	To	Takeoff	Landing	Used	T/o Time	Ld Time	Duration	Ld	Distance	Ht.

Pilot's acceptance prior to today's flight(s): _____

Batteries / **Flight Data**

Cycles this page: / Totals this page:
Cycles brought forward: / Totals brought forward:
Cycles to date: / **Totals to date:**

Discrepancies / Incidents

Flt. No.	Description of Technical Issue	Corrective Action Performed (if required)	Approval R-T-F status Signature / Date

AIRCRAFT DAILY LOG, CONTINUED

Flt. No.	Notes (e.g. purpose of flight, weather, sensors used, camera settings, results, lessons learned)

AIRCRAFT DAILY LOG

DATE:

Flt. No.	P-f. I. / C. Cal.	Crew Names		Location		Fuel / Battery			Flight Data					
		Pilot	Other	From	To	Takeoff	Landing	Used	T/o Time	Ld Time	Duration	Ld	Distance	Ht.

Pilot's acceptance prior to today's flight(s):

Batteries

Cycles this page:

Cycles brought forward:

Cycles to date:

Flight Data

Totals this page:

Totals brought forward:

Totals to date:

Discrepancies / Incidents

Flt. No.	Description of Technical Issue	Corrective Action Performed (if required)	Approval R-T-F status	
			Signature / Date	

AIRCRAFT DAILY LOG, CONTINUED

Flt. No.	Notes (e.g. purpose of flight, weather, sensors used, camera settings, results, lessons learned)

AIRCRAFT DAILY LOG

DATE:

Flt. No.	P-f. I. / C. Cal.	Crew Names		Location		Fuel / Battery			Flight Data					
		Pilot	Other	From	To	Takeoff	Landing	Used	T/o Time	Ld Time	Duration	Ld	Distance	Ht.

Pilot's acceptance prior to today's flight(s): _____

Batteries	Flight Data
Cycles this page:	Totals this page:
Cycles brought forward:	Totals brought forward:
Cycles to date:	**Totals to date:**

Discrepancies / Incidents

Flt. No.	Description of Technical Issue	Corrective Action Performed (if required)	Approval R-T-F status — Signature / Date

AIRCRAFT DAILY LOG, CONTINUED

Flt. No.	Notes (e.g. purpose of flight, weather, sensors used, camera settings, results, lessons learned)

AIRCRAFT DAILY LOG

DATE:

Flt. No.	P-f.l. / C. Cal.	Crew Names		Location		Fuel / Battery			T/o Time	Ld Time	Flight Data			
		Pilot	Other	From	To	Takeoff	Landing	Used			Duration	Ld	Distance	Ht.

Cycles this page:
Cycles brought forward:
Cycles to date:

Totals this page:
Totals brought forward:
Totals to date:

Batteries

Flight Data

Pilot's acceptance prior to today's flight(s): _____

Discrepancies / Incidents

Flt. No.	Description of Technical Issue	Corrective Action Performed (if required)	Approval R-T-F status Signature / Date

Aircraft Daily Log, continued

Flt. No.	Notes (e.g. purpose of flight, weather, sensors used, camera settings, results, lessons learned)

AIRCRAFT DAILY LOG

DATE:

Flt. No.	P-f. I. / C. Cal.	Crew Names		Location		Fuel / Battery			Flight Data						Approval R-T-F status
		Pilot	Other	From	To	Takeoff	Landing	Used	T/o Time	Ld Time	Duration	Ld	Distance	Ht.	

Pilot's acceptance prior to today's flight(s):

Batteries

Cycles this page:

Cycles brought forward:

Cycles to date:

Flight Data

Totals this page:

Totals brought forward:

Totals to date:

Discrepancies / Incidents

Flt. No.	Description of Technical Issue	Corrective Action Performed (if required)	Signature / Date

AIRCRAFT DAILY LOG, CONTINUED

Flt. No.	Notes (e.g. purpose of flight, weather, sensors used, camera settings, results, lessons learned)

AIRCRAFT DAILY LOG

DATE:

66

Flt. No.	P-f. I. / C. Cal.	Crew Names		Location		Fuel / Battery			Flight Data					
		Pilot	Other	From	To	Takeoff	Landing	Used	T/o Time	Ld Time	Duration	Ld	Distance	Ht.

Pilot's acceptance prior to today's flight(s):

Batteries

Cycles this page:
Cycles brought forward:
Cycles to date:

Flight Data

Totals this page:
Totals brought forward:
Totals to date:

Discrepancies / Incidents

Flt. No.	Description of Technical Issue	Corrective Action Performed (if required)	Approval R-T-F status Signature / Date

Aircraft Daily Log, continued

Flt. No.	Notes (e.g. purpose of flight, weather, sensors used, camera settings, results, lessons learned)

AIRCRAFT DAILY LOG

DATE: _____

Flt. No.	P-f.I. / C. Cal.	Crew Names		Location		Fuel / Battery			Flight Data					
		Pilot	Other	From	To	Takeoff	Landing	Used	T/o Time	Ld Time	Duration	Ld	Distance	Ht.

Pilot's acceptance prior to today's flight(s): _____

Batteries

Cycles this page: _____
Cycles brought forward: _____
Cycles to date: _____

Flight Data

Totals this page: _____
Totals brought forward: _____
Totals to date: _____

Discrepancies / Incidents

Flt. No.	Description of Technical Issue	Corrective Action Performed (if required)	Approval R-T-F status	Signature / Date

AIRCRAFT DAILY LOG, CONTINUED

Flt. No.	Notes (e.g. purpose of flight, weather, sensors used, camera settings, results, lessons learned)

AIRCRAFT DAILY LOG

DATE:

Flt. No.	P-f.l. / C. Cal.	Crew Names		Location		Fuel / Battery			Flight Data					
		Pilot	Other	From	To	Takeoff	Landing	Used	T/o Time	Ld Time	Duration	Ld	Distance	Ht.

Batteries

Cycles this page:
Cycles brought forward:
Cycles to date:

Flight Data

Totals this page:
Totals brought forward:
Totals to date:

Pilot's acceptance prior to today's flight(s):

Discrepancies / Incidents

Flt. No.	Description of Technical Issue	Corrective Action Performed (if required)	Approval R-T-F status
			Signature / Date

AIRCRAFT DAILY LOG, CONTINUED

Flt. No.	Notes (e.g. purpose of flight, weather, sensors used, camera settings, results, lessons learned)

Aircraft Daily Log

DATE:

Flt. No.	P-f.I. / C. Cal.	Crew Names		Location		Fuel / Battery			Flight Data						
		Pilot	Other	From	To	Takeoff	Landing	Used	T/o Time	Ld Time	Duration	Ld	Distance	Ht.

Pilot's acceptance prior to today's flight(s):

Batteries

Cycles this page:

Cycles brought forward:

Cycles to date:

Flight Data

Totals this page:

Totals brought forward:

Totals to date:

Discrepancies / Incidents

Flt. No.	Description of Technical Issue	Corrective Action Performed (if required)	Approval R-T-F status
			Signature / Date

Aircraft Daily Log, continued

Flt. No.	Notes (e.g. purpose of flight, weather, sensors used, camera settings, results, lessons learned)

AIRCRAFT DAILY LOG

DATE:

Flt. No.	P-f. I. / C. Cal.	Crew Names			Location		Fuel / Battery			Flight Data						
		Pilot	Other		From	To	Takeoff	Landing	Used	T/o Time	Ld Time	Duration	Ld	Distance	Ht.

Pilot's acceptance prior to today's flight(s): _____

Batteries

Cycles this page: _____
Cycles brought forward: _____
Cycles to date: _____

Flight Data

Totals this page: _____
Totals brought forward: _____
Totals to date: _____

Discrepancies / Incidents

Approval R-T-F status

Flt. No.	Description of Technical Issue	Corrective Action Performed (if required)	Signature / Date

74

AIRCRAFT DAILY LOG, CONTINUED

Flt. No.	Notes (e.g. purpose of flight, weather, sensors used, camera settings, results, lessons learned)

AIRCRAFT DAILY LOG

DATE:

Flt. No.	P-f. I. / C. Cal.	Crew Names		Location		Fuel / Battery			Flight Data					
		Pilot	Other	From	To	Takeoff	Landing	Used	T/o Time	Ld Time	Duration	Ld	Distance	Ht.

Pilot's acceptance prior to today's flight(s):

Batteries

Cycles this page:
Cycles brought forward:
Cycles to date:

Flight Data

Totals this page:
Totals brought forward:
Totals to date:

Discrepancies / Incidents

Flt. No.	Description of Technical Issue	Corrective Action Performed (if required)	Approval R-T-F status
			Signature / Date

AIRCRAFT DAILY LOG, CONTINUED

Flt. No.	Notes (e.g. purpose of flight, weather, sensors used, camera settings, results, lessons learned)

AIRCRAFT DAILY LOG

DATE:

Flt. No.	P-f. l. / C. Cal.	Crew Names		Location		Fuel / Battery			T/o Time	Ld Time	Flight Data			
		Pilot	Other	From	To	Takeoff	Landing	Used			Duration	Ld	Distance	Ht.

Pilot's acceptance prior to today's flight(s): _____

Batteries

Cycles this page: _____
Cycles brought forward: _____
Cycles to date: _____

Flight Data

Totals this page: _____
Totals brought forward: _____
Totals to date: _____

Discrepancies / Incidents

Flt. No.	Description of Technical Issue	Corrective Action Performed (if required)	Approval R-T-F status Signature / Date

AIRCRAFT DAILY LOG, CONTINUED

Flt. No.	Notes (e.g. purpose of flight, weather, sensors used, camera settings, results, lessons learned)

AIRCRAFT DAILY LOG

DATE:

Flt. No.	P-f. I. / C. Cal.	Crew Names		Location		Fuel / Battery			Flight Data						
		Pilot	Other	From	To	Takeoff	Landing	Used	T/o Time	Ld Time	Duration	Ld	Distance	Ht.

Batteries
Cycles this page:
Cycles brought forward:
Cycles to date:

Flight Data
Totals this page:
Totals brought forward:
Totals to date:

Pilot's acceptance
prior to today's flight(s): _____

Discrepancies / Incidents

Flt. No.	Description of Technical Issue	Corrective Action Performed (if required)	Approval R-T-F status
			Signature / Date

Aircraft Daily Log, continued

Flt. No.	Notes (e.g. purpose of flight, weather, sensors used, camera settings, results, lessons learned)

AIRCRAFT DAILY LOG

DATE: _____

Flt. No.	P-f. I. / C. Cal.	Crew Names		Location		Fuel / Battery			Flight Data						
		Pilot	Other	From	To	Takeoff	Landing	Used	T/o Time	Ld Time	Duration	Ld	Distance	Ht.

Pilot's acceptance prior to today's flight(s): _____

Batteries
Cycles this page: _____
Cycles brought forward: _____
Cycles to date: _____

Flight Data
Totals this page: _____
Totals brought forward: _____
Totals to date: _____

Discrepancies / Incidents

Flt. No.	Description of Technical Issue	Corrective Action Performed (if required)	Approval R-T-F status
			Signature / Date

AIRCRAFT DAILY LOG, CONTINUED

Flt. No.	Notes (e.g. purpose of flight, weather, sensors used, camera settings, results, lessons learned)

AIRCRAFT DAILY LOG

DATE: _____

Flt. No.	P-f. I. / C. Cal.	Crew Names		Location		Fuel / Battery			T/o Time	Ld Time	Flight Data			
		Pilot	Other	From	To	Takeoff	Landing	Used			Duration	Ld	Distance	Ht.

Pilot's acceptance prior to today's flight(s): _____

Flight Data / **Batteries**

- Totals this page: _____ Cycles this page: _____
- Totals brought forward: _____ Cycles brought forward: _____
- **Totals to date:** _____ **Cycles to date:** _____

Discrepancies / Incidents

Flt. No.	Description of Technical Issue	Corrective Action Performed (if required)	Approval R-T-F status / Signature / Date

AIRCRAFT DAILY LOG, CONTINUED

Flt. No.	Notes (e.g. purpose of flight, weather, sensors used, camera settings, results, lessons learned)

AIRCRAFT DAILY LOG

DATE: _____

Flt. No.	P-f. I. / C. Cal.	Crew Names		Location		Fuel / Battery			Flight Data					
		Pilot	Other	From	To	Takeoff	Landing	Used	T/o Time	Ld Time	Duration	Ld	Distance	Ht.

Pilot's acceptance prior to today's flight(s): _____

Batteries

Cycles this page:	Totals this page:
Cycles brought forward:	Totals brought forward:
Cycles to date:	**Totals to date:**

Flight Data

Discrepancies / Incidents

Flt. No.	Description of Technical Issue	Corrective Action Performed (if required)	Approval R-T-F status
			Signature / Date

AIRCRAFT DAILY LOG, CONTINUED

Flt. No.	Notes (e.g. purpose of flight, weather, sensors used, camera settings, results, lessons learned)

AIRCRAFT DAILY LOG

DATE:

Flt. No.	P-f. I. / C. Cal.	Crew Names		Location		Fuel / Battery			Flight Data					
		Pilot	Other	From	To	Takeoff	Landing	Used	T/o Time	Ld Time	Duration	Ld	Distance	Ht.

Pilot's acceptance prior to today's flight(s):

Batteries

Cycles this page:

Cycles brought forward:

Cycles to date:

Flight Data

Totals this page:

Totals brought forward:

Totals to date:

Discrepancies / Incidents

Flt. No.	Description of Technical Issue	Corrective Action Performed (if required)	Approval R-T-F status	Signature / Date

88

Aircraft Daily Log, continued

Flt. No.	Notes (e.g. purpose of flight, weather, sensors used, camera settings, results, lessons learned)

AIRCRAFT DAILY LOG

DATE:

Flt. No.	P-f.I. / C. Cal.	Crew Names		Location		Fuel / Battery			T/o Time	Ld Time	Flight Data			
		Pilot	Other	From	To	Takeoff	Landing	Used			Duration	Ld	Distance	Ht.

Pilot's acceptance prior to today's flight(s):

Batteries

Cycles this page:
Cycles brought forward:
Cycles to date:

Flight Data

Totals this page:
Totals brought forward:
Totals to date:

Discrepancies / Incidents

Flt. No.	Description of Technical Issue	Corrective Action Performed (if required)	Approval R-T-F status Signature / Date

AIRCRAFT DAILY LOG, CONTINUED

Flt. No.	Notes (e.g. purpose of flight, weather, sensors used, camera settings, results, lessons learned)

AIRCRAFT DAILY LOG

DATE:

Flt. No.	P-f. I. / C. Cal.	Crew Names		Location		Fuel / Battery			Flight Data					
		Pilot	Other	From	To	Takeoff	Landing	Used	T/o Time	Ld Time	Duration	Ld	Distance	Ht.

Batteries

Cycles this page: _____

Cycles brought forward: _____

Cycles to date: _____

Flight Data

Totals this page: _____

Totals brought forward: _____

Totals to date: _____

Pilot's acceptance prior to today's flight(s): _____

Discrepancies / Incidents

Flt. No.	Description of Technical Issue	Corrective Action Performed (if required)	Approval R-T-F status	Signature / Date

AIRCRAFT DAILY LOG, CONTINUED

Flt. No.	Notes (e.g. purpose of flight, weather, sensors used, camera settings, results, lessons learned)

AIRCRAFT DAILY LOG   DATE:

Flt. No.	P-f. I. / C. Cal.	Crew Names		Location		Fuel / Battery			Flight Data					
		Pilot	Other	From	To	Takeoff	Landing	Used	T/o Time	Ld Time	Duration	Ld	Distance	Ht.
											Totals this page:			
											Totals brought forward:			
											Totals to date:			

Flight Data

Batteries

Pilot's acceptance prior to today's flight(s):

Cycles this page:
Cycles brought forward:
Cycles to date: _____

Discrepancies / Incidents

Flt. No.	Description of Technical Issue	Corrective Action Performed (if required)	Approval R-T-F status
			Signature / Date

AIRCRAFT DAILY LOG, CONTINUED

Flt. No.	Notes (e.g. purpose of flight, weather, sensors used, camera settings, results, lessons learned)

AIRCRAFT DAILY LOG

DATE:

Flt. No.	P-f. I. / C. Cal.	Crew Names		Location		Fuel / Battery			Flight Data					
		Pilot	Other	From	To	Takeoff	Landing	Used	T/o Time	Ld Time	Duration	Ld	Distance	Ht.
									Totals this page:					
									Totals brought forward:					
									Totals to date:					

Pilot's acceptance prior to today's flight(s):

Batteries

Cycles this page:

Cycles brought forward:

Cycles to date:

Flight Data

Discrepancies / Incidents

Approval R-T-F status

Flt. No.	Description of Technical Issue	Corrective Action Performed (if required)	Signature / Date

AIRCRAFT DAILY LOG, CONTINUED

Flt. No.	Notes (e.g. purpose of flight, weather, sensors used, camera settings, results, lessons learned)

AIRCRAFT DAILY LOG

DATE:

Flt. No.	P-f. I. / C. Cal.	Crew Names		Location		Fuel / Battery			Flight Data						
		Pilot	Other	From	To	Takeoff	Landing	Used	T/o Time	Ld Time	Duration	Ld	Distance	Ht.

Pilot's acceptance prior to today's flight(s):

Batteries

Cycles this page:
Cycles brought forward:
Cycles to date:

Flight Data

Totals this page:
Totals brought forward:
Totals to date:

Discrepancies / Incidents

Flt. No.	Description of Technical Issue	Corrective Action Performed (if required)	Approval R-T-F status
			Signature / Date

AIRCRAFT DAILY LOG, CONTINUED

Flt. No.	Notes (e.g. purpose of flight, weather, sensors used, camera settings, results, lessons learned)

AIRCRAFT DAILY LOG   DATE:

Flt. No.	P-f. I. / C. Cal.	Crew Names		Location		Fuel / Battery			Flight Data						
		Pilot	Other	From	To	Takeoff	Landing	Used	T/o Time	Ld Time	Duration	Ld	Distance	Ht.

Batteries / **Flight Data**

Cycles this page:	Totals this page:
Cycles brought forward:	Totals brought forward:
Cycles to date:	**Totals to date:**

Pilot's acceptance prior to today's flight(s): _____

Approval R-T-F status — Signature / Date

Discrepancies / Incidents

Flt. No.	Description of Technical Issue	Corrective Action Performed (if required)	Signature / Date

Flt. No.	Notes (e.g. purpose of flight, weather, sensors used, camera settings, results, lessons learned)

AIRCRAFT DAILY LOG

DATE:

Flt. No.	P-f.l. / C. Cal.	Crew Names		Location		Fuel / Battery			T/o Time	Ld Time	Flight Data			
		Pilot	Other	From	To	Takeoff	Landing	Used			Duration	Ld	Distance	Ht.

Pilot's acceptance prior to today's flight(s):

Batteries
Cycles this page:
Cycles brought forward:
Cycles to date:

Flight Data
Totals this page:
Totals brought forward:
Totals to date:

Discrepancies / Incidents

Flt. No.	Description of Technical Issue	Corrective Action Performed (if required)	Approval R-T-F status
			Signature / Date

AIRCRAFT DAILY LOG, CONTINUED

Flt. No.	Notes (e.g. purpose of flight, weather, sensors used, camera settings, results, lessons learned)

AIRCRAFT DAILY LOG

DATE:

Flt. No.	P-f. I. / C. Cal.	Crew Names		Location		Fuel / Battery			Flight Data					
		Pilot	Other	From	To	Takeoff	Landing	Used	T/o Time	Ld Time	Duration	Ld	Distance	Ht.

Pilot's acceptance prior to today's flight(s):

Batteries

Cycles this page:

Cycles brought forward:

Cycles to date:

Flight Data

Totals this page:

Totals brought forward:

Totals to date:

Approval R-T-F status

Signature / Date

Discrepancies / Incidents

Flt. No.	Description of Technical Issue	Corrective Action Performed (if required)

AIRCRAFT DAILY LOG, CONTINUED

Flt. No.	Notes (e.g. purpose of flight, weather, sensors used, camera settings, results, lessons learned)

AIRCRAFT DAILY LOG

DATE:

Flt. No.	P-f. I. / C. Cal.	Crew Names		Location		Fuel / Battery			T/o Time	Ld Time	Flight Data			
		Pilot	Other	From	To	Takeoff	Landing	Used			Duration	Ld	Distance	Ht.

Pilot's acceptance prior to today's flight(s):

Cycles this page:
Cycles brought forward:
Cycles to date:

Batteries

Flight Data

Totals this page:
Totals brought forward:
Totals to date:

Discrepancies / Incidents

Flt. No.	Description of Technical Issue	Corrective Action Performed (if required)	Approval R-T-F status
			Signature / Date

Aircraft Daily Log, continued

Flt. No.	Notes (e.g. purpose of flight, weather, sensors used, camera settings, results, lessons learned)

AIRCRAFT DAILY LOG

DATE: _____

Flt. No.	P-f.I. / C. Cal.	Crew Names		Location		Fuel / Battery			Flight Data					
		Pilot	Other	From	To	Takeoff	Landing	Used	T/o Time	Ld Time	Duration	Ld	Distance	Ht.

Pilot's acceptance prior to today's flight(s): _____

Batteries

Cycles this page:
Cycles brought forward:
Cycles to date:

Flight Data

Totals this page:
Totals brought forward:
Totals to date:

Discrepancies / Incidents

Approval R-T-F status

Flt. No.	Description of Technical Issue	Corrective Action Performed (if required)	Signature / Date

AIRCRAFT DAILY LOG, CONTINUED

Flt. No.	Notes (e.g. purpose of flight, weather, sensors used, camera settings, results, lessons learned)

AIRCRAFT DAILY LOG

DATE:

Flt. No.	P-f.I. / C. Cal.	Crew Names		Location		Fuel / Battery			Flight Data						
		Pilot	Other	From	To	Takeoff	Landing	Used	T/o Time	Ld Time	Duration	Ld	Distance	Ht.	

Pilot's acceptance prior to today's flight(s):

Batteries

Cycles this page:
Cycles brought forward:
Cycles to date:

Flight Data

Totals this page:
Totals brought forward:
Totals to date:

Approval R-T-F status

Signature / Date

Discrepancies / Incidents

Flt. No.	Description of Technical Issue	Corrective Action Performed (if required)

110

AIRCRAFT DAILY LOG, CONTINUED

Flt. No.	Notes (e.g. purpose of flight, weather, sensors used, camera settings, results, lessons learned)

AIRCRAFT DAILY LOG

DATE: _____

Flt. No.	P-f. I. / C. Cal.	Crew Names		Location		Fuel / Battery			Flight Data					
		Pilot	Other	From	To	Takeoff	Landing	Used	T/o Time	Ld Time	Duration	Ld	Distance	Ht.

Pilot's acceptance prior to today's flight(s): _____

Batteries
Cycles this page:
Cycles brought forward:
Cycles to date:

Flight Data
Totals this page:
Totals brought forward:
Totals to date:

Discrepancies / Incidents

Flt. No.	Description of Technical Issue	Corrective Action Performed (if required)	Approval R-T-F status
			Signature / Date

112

AIRCRAFT DAILY LOG, CONTINUED

Flt. No.	Notes (e.g. purpose of flight, weather, sensors used, camera settings, results, lessons learned)

AIRCRAFT DAILY LOG

DATE:

Flt. No.	P-f.l. / C. Cal.	Crew Names		Location		Fuel / Battery			Flight Data					
		Pilot	Other	From	To	Takeoff	Landing	Used	T/o Time	Ld Time	Duration	Ld	Distance	Ht.

Pilot's acceptance prior to today's flight(s):

Batteries

Cycles this page: _____
Cycles brought forward: _____
Cycles to date: _____

Flight Data

Totals this page: _____
Totals brought forward: _____
Totals to date: _____

Discrepancies / Incidents

Flt. No.	Description of Technical Issue	Corrective Action Performed (if required)	Approval R-T-F status Signature / Date

114

AIRCRAFT DAILY LOG, CONTINUED

Flt. No.	Notes (e.g. purpose of flight, weather, sensors used, camera settings, results, lessons learned)

AIRCRAFT DAILY LOG

DATE:

Flt. No.	P.f. I. / C. Cal.	Crew Names		Location		Fuel / Battery			Flight Data						
		Pilot	Other	From	To	Takeoff	Landing	Used	T/o Time	Ld Time	Duration	Ld	Distance	Ht.

Pilot's acceptance prior to today's flight(s):

Batteries

Cycles this page:
Cycles brought forward:
Cycles to date:

Flight Data

Totals this page:
Totals brought forward:
Totals to date:

Approval R-T-F status

Signature / Date

Discrepancies / Incidents

Flt. No.	Description of Technical Issue	Corrective Action Performed (if required)

Aircraft Daily Log, continued

Flt. No.	Notes (e.g. purpose of flight, weather, sensors used, camera settings, results, lessons learned)

AIRCRAFT DAILY LOG

DATE:

| Flt. No. | P-f. I. / C. Cal. | Crew Names | | Location | | Fuel / Battery | | | | | Flight Data | | | | |
		Pilot	Other	From	To	Takeoff	Landing	Used	T/o Time	Ld Time	Duration	Ld	Distance	Ht.

Pilot's acceptance prior to today's flight(s):

Batteries

Cycles this page:

Cycles brought forward:

Cycles to date:

Flight Data

Totals this page:

Totals brought forward:

Totals to date:

Discrepancies / Incidents

| Flt. No. | Description of Technical Issue | Corrective Action Performed (if required) | Approval R-T-F status |
			Signature / Date

Aircraft Daily Log, continued

Flt. No.	Notes (e.g. purpose of flight, weather, sensors used, camera settings, results, lessons learned)

AIRCRAFT DAILY LOG

DATE:

Flt. No.	P-f. I. / C. Cal.	Crew Names		Location		Fuel / Battery			Flight Data					
		Pilot	Other	From	To	Takeoff	Landing	Used	T/o Time	Ld Time	Duration	Ld	Distance	Ht.

Batteries

Cycles this page:

Cycles brought forward:

Cycles to date:

Pilot's acceptance prior to today's flight(s): _____

Flight Data

Totals this page:

Totals brought forward:

Totals to date:

Discrepancies / Incidents

Flt. No.	Description of Technical Issue	Corrective Action Performed (if required)	Approval R-T-F status Signature / Date

AIRCRAFT DAILY LOG, CONTINUED

Flt. No.	Notes (e.g. purpose of flight, weather, sensors used, camera settings, results, lessons learned)

AIRCRAFT DAILY LOG

DATE:

Flt. No.	P-f. I. / C. Cal.	Crew Names		Location		Fuel / Battery			Flight Data					
		Pilot	Other	From	To	Takeoff	Landing	Used	T/o Time	Ld Time	Duration	Ld	Distance	Ht.

Pilot's acceptance prior to today's flight(s):

Batteries
Cycles this page:
Cycles brought forward:
Cycles to date:

Flight Data
Totals this page:
Totals brought forward:
Totals to date:

Discrepancies / Incidents

Flt. No.	Description of Technical Issue	Corrective Action Performed (if required)	Approval R-T-F status — Signature / Date

AIRCRAFT DAILY LOG, CONTINUED

Flt. No.	Notes (e.g. purpose of flight, weather, sensors used, camera settings, results, lessons learned)

AIRCRAFT DAILY LOG

DATE:

Flt. No.	P-f.I. / C. Cal.	Crew Names		Location		Fuel / Battery			Flight Data						Approval R-T-F status
		Pilot	Other	From	To	Takeoff	Landing	Used	T/o Time	Ld Time	Duration	Ld	Distance	Ht.	Signature / Date

Pilot's acceptance prior to today's flight(s):

Batteries
Cycles this page:
Cycles brought forward:
Cycles to date:

Flight Data
Totals this page:
Totals brought forward:
Totals to date:

Discrepancies / Incidents

Flt. No.	Description of Technical Issue	Corrective Action Performed (if required)

AIRCRAFT DAILY LOG, CONTINUED

Flt. No.	Notes (e.g. purpose of flight, weather, sensors used, camera settings, results, lessons learned)

AIRCRAFT DAILY LOG DATE:

Flt. No.	P-f.I. / C. Cal.	Crew Names		Location		Fuel / Battery			T/o Time	Ld Time	Flight Data			
		Pilot	Other	From	To	Takeoff	Landing	Used			Duration	Ld	Distance	Ht.

Pilot's acceptance prior to today's flight(s):

Batteries

Cycles this page:
Cycles brought forward:
Cycles to date:

Flight Data
Totals this page:
Totals brought forward:
Totals to date:

Discrepancies / Incidents

Flt. No.	Description of Technical Issue	Corrective Action Performed (if required)	Approval R-T-F status
			Signature / Date

AIRCRAFT DAILY LOG, CONTINUED

Flt. No.	Notes (e.g. purpose of flight, weather, sensors used, camera settings, results, lessons learned)

AIRCRAFT DAILY LOG

DATE:

Flt. No.	P-f.I. / C. Cal.	Crew Names		Location		Fuel / Battery			T/o Time	Ld Time	Flight Data			
		Pilot	Other	From	To	Takeoff	Landing	Used			Duration	Ld	Distance	Ht.

Pilot's acceptance prior to today's flight(s): _____

Batteries
- Cycles this page:
- Cycles brought forward:
- **Cycles to date:**

Flight Data
- Totals this page:
- Totals brought forward:
- **Totals to date:**

Discrepancies / Incidents

Flt. No.	Description of Technical Issue	Corrective Action Performed (if required)	Approval R-T-F status Signature / Date

AIRCRAFT DAILY LOG, CONTINUED

Flt. No.	Notes (e.g. purpose of flight, weather, sensors used, camera settings, results, lessons learned)

AIRCRAFT DAILY LOG

DATE:

Flt. No.	P-f. I. / C. Cal.	Crew Names		Location		Fuel / Battery					Flight Data			
		Pilot	Other	From	To	Takeoff	Landing	Used	T/o Time	Ld Time	Duration	Ld	Distance	Ht.

Pilot's acceptance prior to today's flight(s): _____

Batteries

Cycles this page: _____
Cycles brought forward: _____
Cycles to date: _____

Flight Data

Totals this page: _____
Totals brought forward: _____
Totals to date: _____

Discrepancies / Incidents

Flt. No.	Description of Technical Issue	Corrective Action Performed (if required)	Approval R-T-F status Signature / Date

Aircraft Daily Log, continued

Flt. No.	Notes (e.g. purpose of flight, weather, sensors used, camera settings, results, lessons learned)

AIRCRAFT DAILY LOG

DATE:

Flt. No.	P-f.I. / C. Cal.	Crew Names		Location		Fuel / Battery			Flight Data					
		Pilot	Other	From	To	Takeoff	Landing	Used	T/o Time	Ld Time	Duration	Ld	Distance	Ht.

Pilot's acceptance prior to today's flight(s): _____

Batteries — Cycles this page: / Cycles brought forward: / **Cycles to date:**

Flight Data — Totals this page: / Totals brought forward: / **Totals to date:**

Discrepancies / Incidents

Flt. No.	Description of Technical Issue	Corrective Action Performed (if required)	Approval R-T-F status Signature / Date

AIRCRAFT DAILY LOG, CONTINUED

Flt. No.	Notes (e.g. purpose of flight, weather, sensors used, camera settings, results, lessons learned)

AIRCRAFT DAILY LOG

DATE:

Flt. No.	P-f. l. / C. Cal.	Crew Names		Location		Fuel / Battery			Flight Data						Approval R-T-F status
		Pilot	Other	From	To	Takeoff	Landing	Used	T/o Time	Ld Time	Duration	Ld	Distance	Ht.	Signature / Date

Pilot's acceptance prior to today's flight(s):

Batteries — Cycles this page: / Cycles brought forward: / **Cycles to date:**

Flight Data — Totals this page: / Totals brought forward: / **Totals to date:**

Discrepancies / Incidents

Flt. No.	Description of Technical Issue	Corrective Action Performed (if required)

AIRCRAFT DAILY LOG, CONTINUED

Flt. No.	Notes (e.g. purpose of flight, weather, sensors used, camera settings, results, lessons learned)

Aircraft Daily Log

DATE:

Flt. No.	P-f. I. / C. Cal.	Crew Names		Location		Fuel / Battery			Flight Data						
		Pilot	Other	From	To	Takeoff	Landing	Used	T/o Time	Ld Time	Duration	Ld	Distance	Ht.

Pilot's acceptance prior to today's flight(s):

Batteries

Cycles this page:
Cycles brought forward:
Cycles to date:

Flight Data

Totals this page:
Totals brought forward:
Totals to date:

Discrepancies / Incidents

Flt. No.	Description of Technical Issue	Corrective Action Performed (if required)	Approval R-T-F status
			Signature / Date

AIRCRAFT DAILY LOG, CONTINUED

Flt. No.	Notes (e.g. purpose of flight, weather, sensors used, camera settings, results, lessons learned)

AIRCRAFT DAILY LOG

DATE: _____

Flt. No.	P-f. I. / C. Cal.	Crew Names		Location		Fuel / Battery			Flight Data						
		Pilot	Other	From	To	Takeoff	Landing	Used	T/o Time	Ld Time	Duration	Ld	Distance	Ht.

Flight Data

Totals this page: _____

Totals brought forward: _____

Totals to date: _____

Batteries

Cycles this page: _____

Cycles brought forward: _____

Cycles to date: _____

Pilot's acceptance prior to today's flight(s):

_____

Discrepancies / Incidents

Approval R-T-F status

Flt. No.	Description of Technical Issue	Corrective Action Performed (if required)	Signature / Date

Aircraft Daily Log, continued

Flt. No.	Notes (e.g. purpose of flight, weather, sensors used, camera settings, results, lessons learned)

AIRCRAFT DAILY LOG

DATE:

Flt. No.	P-f.l. / C. Cal.	Crew Names		Location		Fuel / Battery			T/o Time	Ld Time	Flight Data			
		Pilot	Other	From	To	Takeoff	Landing	Used			Duration	Ld	Distance	Ht.

Cycles this page:

Cycles brought forward:

Cycles to date:

Batteries

Pilot's acceptance prior to today's flight(s):

_____

Flight Data

Totals this page:

Totals brought forward:

Totals to date:

Discrepancies / Incidents

Flt. No.	Description of Technical Issue	Corrective Action Performed (if required)	Approval R-T-F status
			Signature / Date

AIRCRAFT DAILY LOG, CONTINUED

Flt. No.	Notes (e.g. purpose of flight, weather, sensors used, camera settings, results, lessons learned)

Aircraft Daily Log

DATE: _____

Flt. No.	P-f.I. / C. Cal.	Crew Names		Location		Fuel / Battery					Flight Data				
		Pilot	Other	From	To	Takeoff	Landing	Used	T/o Time	Ld Time	Duration	Ld	Distance	Ht.

Pilot's acceptance prior to today's flight(s): _____

Batteries

Cycles this page: _____
Cycles brought forward: _____
Cycles to date: _____

Flight Data

Totals this page: _____
Totals brought forward: _____
Totals to date: _____

Discrepancies / Incidents

Approval R-T-F status

Flt. No.	Description of Technical Issue	Corrective Action Performed (if required)	Signature / Date

AIRCRAFT DAILY LOG, CONTINUED

Flt. No.	Notes (e.g. purpose of flight, weather, sensors used, camera settings, results, lessons learned)

AIRCRAFT DAILY LOG

DATE: _____

Flt. No.	P-f.I. / C. Cal.	Crew Names		Location		Fuel / Battery			Flight Data					
		Pilot	Other	From	To	Takeoff	Landing	Used	T/o Time	Ld Time	Duration	Ld	Distance	Ht.

Pilot's acceptance prior to today's flight(s): _____

Batteries

Cycles this page: _____
Cycles brought forward: _____
Cycles to date: _____

Flight Data

Totals this page: _____
Totals brought forward: _____
Totals to date: _____

Discrepancies / Incidents

Flt. No.	Description of Technical Issue	Corrective Action Performed (if required)	Approval R-T-F status
			Signature / Date

Flt. No.	Notes (e.g. purpose of flight, weather, sensors used, camera settings, results, lessons learned)

AIRCRAFT DAILY LOG

DATE:

Flt. No.	P-f. I. / C. Cal.	Crew Names		Location		Fuel / Battery					Flight Data			
		Pilot	Other	From	To	Takeoff	Landing	Used	T/o Time	Ld Time	Duration	Ld	Distance	Ht.

Pilot's acceptance prior to today's flight(s): _____

Batteries — Cycles this page: _____
Cycles brought forward: _____
Cycles to date: _____

Flight Data — Totals this page: _____
Totals brought forward: _____
Totals to date: _____

Discrepancies / Incidents

Approval R-T-F status

Flt. No.	Description of Technical Issue	Corrective Action Performed (if required)	Signature / Date

Aircraft Daily Log, continued

Flt. No.	Notes (e.g. purpose of flight, weather, sensors used, camera settings, results, lessons learned)

AIRCRAFT DAILY LOG

DATE:

Flt. No.	P-f. I. / C. Cal.	Crew Names		Location		Fuel / Battery			T/o Time	Ld Time	Flight Data			
		Pilot	Other	From	To	Takeoff	Landing	Used			Duration	Ld	Distance	Ht.

Pilot's acceptance prior to today's flight(s):

Batteries

Cycles this page:
Cycles brought forward:
Cycles to date:

Flight Data

Totals this page:
Totals brought forward:
Totals to date:

Discrepancies / Incidents

Flt. No.	Description of Technical Issue	Corrective Action Performed (if required)	Approval R-T-F status
			Signature / Date

AIRCRAFT DAILY LOG, CONTINUED

Flt. No.	Notes (e.g. purpose of flight, weather, sensors used, camera settings, results, lessons learned)

AIRCRAFT DAILY LOG

DATE:

Flt. No.	P-f. I. / C. Cal.	Crew Names		Location		Fuel / Battery			Flight Data						
		Pilot	Other	From	To	Takeoff	Landing	Used	T/o Time	Ld Time	Duration	Ld	Distance	Ht.
														✕

Pilot's acceptance prior to today's flight(s):

Batteries

Cycles this page:
Cycles brought forward:
Cycles to date:

Flight Data

Totals this page:
Totals brought forward:
Totals to date:

Discrepancies / Incidents

Flt. No.	Description of Technical Issue	Corrective Action Performed (if required)	Approval R-T-F status	Signature / Date

Aircraft Daily Log, continued

Flt. No.	Notes (e.g. purpose of flight, weather, sensors used, camera settings, results, lessons learned)

AIRCRAFT DAILY LOG

DATE: _____

Flt. No.	P-f. I. / C. Cal.	Crew Names		Location		Fuel / Battery			Flight Data						
		Pilot	Other	From	To	Takeoff	Landing	Used	T/o Time	Ld Time	Duration	Ld	Distance	Ht.

Batteries / Flight Data

Cycles this page: _____ Totals this page: _____

Cycles brought forward: _____ Totals brought forward: _____

Cycles to date: _____ **Totals to date:** _____

Pilot's acceptance prior to today's flight(s): _____

Discrepancies / Incidents

Flt. No.	Description of Technical Issue	Corrective Action Performed (if required)	Approval R-T-F status
			Signature / Date

Flt. No.	Notes (e.g. purpose of flight, weather, sensors used, camera settings, results, lessons learned)

AIRCRAFT DAILY LOG

DATE:

Flt. No.	P-f. I. / C. Cal.	Crew Names		Location		Fuel / Battery					Flight Data				
		Pilot	Other	From	To	Takeoff	Landing	Used	T/o Time	Ld Time	Duration	Ld	Distance	Ht.

Batteries

Cycles this page:
Cycles brought forward:
Cycles to date:

Flight Data

Totals this page:
Totals brought forward:
Totals to date:

Pilot's acceptance prior to today's flight(s):

Discrepancies / Incidents

Flt. No.	Description of Technical Issue	Corrective Action Performed (if required)	Approval R-T-F status	Signature / Date

AIRCRAFT DAILY LOG, CONTINUED

Flt. No.	Notes (e.g. purpose of flight, weather, sensors used, camera settings, results, lessons learned)

AIRCRAFT DAILY LOG

DATE:

Flt. No.	P-f. I. / C. Cal.	Crew Names		Location		Fuel / Battery			Flight Data					
		Pilot	Other	From	To	Takeoff	Landing	Used	T/o Time	Ld Time	Duration	Ld	Distance	Ht.

Pilot's acceptance prior to today's flight(s):

Batteries

Cycles this page:

Cycles brought forward:

Cycles to date:

Flight Data

Totals this page:

Totals brought forward:

Totals to date:

Discrepancies / Incidents

Flt. No.	Description of Technical Issue	Corrective Action Performed (if required)	Approval R-T-F status
			Signature / Date

AIRCRAFT DAILY LOG, CONTINUED

Flt. No.	Notes (e.g. purpose of flight, weather, sensors used, camera settings, results, lessons learned)

AIRCRAFT DAILY LOG

DATE: _____

Flt. No.	P-f. I. / C. Cal.	Crew Names		Location		Fuel / Battery			Flight Data					
		Pilot	Other	From	To	Takeoff	Landing	Used	T/o Time	Ld Time	Duration	Ld	Distance	Ht.

Pilot's acceptance prior to today's flight(s): _____

Batteries

Cycles this page: _____
Cycles brought forward: _____
Cycles to date: _____

Flight Data

Totals this page: _____
Totals brought forward: _____
Totals to date: _____

Discrepancies / Incidents

Flt. No.	Description of Technical Issue	Corrective Action Performed (if required)	Approval R-T-F status
			Signature / Date

Flt. No.	Notes (e.g. purpose of flight, weather, sensors used, camera settings, results, lessons learned)

AIRCRAFT DAILY LOG

DATE:

Flt. No.	P-f. I. / C. Cal.	Crew Names		Location		Fuel / Battery			Flight Data					
		Pilot	Other	From	To	Takeoff	Landing	Used	T/o Time	Ld Time	Duration	Ld	Distance	Ht.

Pilot's acceptance prior to today's flight(s):

Batteries

Cycles this page:
Cycles brought forward:
Cycles to date:

Flight Data

Totals this page:
Totals brought forward:
Totals to date:

Approval R-T-F status

Signature / Date

Discrepancies / Incidents

Flt. No.	Description of Technical Issue	Corrective Action Performed (if required)

AIRCRAFT DAILY LOG, CONTINUED

Flt. No.	Notes (e.g. purpose of flight, weather, sensors used, camera settings, results, lessons learned)

AIRCRAFT DAILY LOG

DATE: _____

Flt. No.	P-f. I. / C. Cal.	Crew Names		Location		Fuel / Battery			T/o Time	Ld Time	Flight Data			
		Pilot	Other	From	To	Takeoff	Landing	Used			Duration	Ld	Distance	Ht.

Pilot's acceptance prior to today's flight(s): _____

Batteries
- Cycles this page: _____
- Cycles brought forward: _____
- **Cycles to date:** _____

Flight Data
- Totals this page: _____
- Totals brought forward: _____
- **Totals to date:** _____

Discrepancies / Incidents

Flt. No.	Description of Technical Issue	Corrective Action Performed (if required)	Approval R-T-F status — Signature / Date

Aircraft Daily Log, continued

Flt. No.	Notes (e.g. purpose of flight, weather, sensors used, camera settings, results, lessons learned)

AIRCRAFT DAILY LOG

DATE: _____

Flt. No.	P-f. I. / C. Cal.	Crew Names		Location		Fuel / Battery			Flight Data						
		Pilot	Other	From	To	Takeoff	Landing	Used	T/o Time	Ld Time	Duration	Ld	Distance	Ht.

Pilot's acceptance prior to today's flight(s): _____

Batteries

Cycles this page: _____
Cycles brought forward: _____
Cycles to date: _____

Flight Data

Totals this page: _____
Totals brought forward: _____
Totals to date: _____

Discrepancies / Incidents

Flt. No.	Description of Technical Issue	Corrective Action Performed (if required)	Approval R-T-F status
			Signature / Date

Flt. No.	Notes (e.g. purpose of flight, weather, sensors used, camera settings, results, lessons learned)

Aircraft Daily Log

DATE:

Flt. No.	P-f. I. / C. Cal.	Crew Names		Location		Fuel / Battery			Flight Data					
		Pilot	Other	From	To	Takeoff	Landing	Used	T/o Time	Ld Time	Duration	Ld	Distance	Ht.

Pilot's acceptance prior to today's flight(s):

Batteries

Cycles this page:
Cycles brought forward:
Cycles to date:

Flight Data

Totals this page:
Totals brought forward:
Totals to date:

Discrepancies / Incidents

Flt. No.	Description of Technical Issue	Corrective Action Performed (if required)	Approval R-T-F status
			Signature / Date

AIRCRAFT DAILY LOG, CONTINUED

Flt. No.	Notes (e.g. purpose of flight, weather, sensors used, camera settings, results, lessons learned)

AIRCRAFT DAILY LOG

DATE: _____

Flt. No.	P-f.I. / C. Cal.	Crew Names		Location		Fuel / Battery			Flight Data					
		Pilot	Other	From	To	Takeoff	Landing	Used	T/o Time	Ld Time	Duration	Ld	Distance	Ht.

Batteries

Cycles this page: _____   Flight Data   Totals this page: _____

Cycles brought forward: _____   Totals brought forward: _____

Cycles to date: _____   **Totals to date:** _____

Pilot's acceptance prior to today's flight(s): _____

Discrepancies / Incidents

Flt. No.	Description of Technical Issue	Corrective Action Performed (if required)	Approval R-T-F status
			Signature / Date

Aircraft Daily Log, continued

Flt. No.	Notes (e.g. purpose of flight, weather, sensors used, camera settings, results, lessons learned)

AIRCRAFT DAILY LOG

DATE:

Flt. No.	P-f. I. / C. Cal.	Crew Names		Location		Fuel / Battery			Flight Data					
		Pilot	Other	From	To	Takeoff	Landing	Used	T/o Time	Ld Time	Duration	Ld	Distance	Ht.

Pilot's acceptance prior to today's flight(s):

Batteries
Cycles this page:
Cycles brought forward:
Cycles to date:

Flight Data
Totals this page:
Totals brought forward:
Totals to date:

Discrepancies / Incidents

Flt. No.	Description of Technical Issue	Corrective Action Performed (if required)	Approval R-T-F status
			Signature / Date

170

Flt. No.	Notes (e.g. purpose of flight, weather, sensors used, camera settings, results, lessons learned)

AIRCRAFT DAILY LOG

DATE:

Flt. No.	P-f. I. / C. Cal.	Crew Names		Location		Fuel / Battery			Flight Data					
		Pilot	Other	From	To	Takeoff	Landing	Used	T/o Time	Ld Time	Duration	Ld	Distance	Ht.

Pilot's acceptance prior to today's flight(s): _____

Batteries

Cycles this page:

Cycles brought forward:

Cycles to date:

Flight Data

Totals this page:

Totals brought forward:

Totals to date:

Discrepancies / Incidents

Approval R-T-F status

Flt. No.	Description of Technical Issue	Corrective Action Performed (if required)	Signature / Date

AIRCRAFT DAILY LOG, CONTINUED

Flt. No.	Notes (e.g. purpose of flight, weather, sensors used, camera settings, results, lessons learned)

AIRCRAFT DAILY LOG

DATE: _____

Flt. No.	P-f.I. / C. Cal.	Crew Names		Location		Fuel / Battery			Flight Data						
		Pilot	Other	From	To	Takeoff	Landing	Used	T/o Time	Ld Time	Duration	Ld	Distance	Ht.

Pilot's acceptance prior to today's flight(s): _____

Batteries

		Cycles this page:		Flight Data			Totals this page:			
Cycles brought forward:					Totals brought forward:		
Cycles to date:					**Totals to date:**		

Discrepancies / Incidents

Flt. No.	Description of Technical Issue	Corrective Action Performed (if required)	Approval R-T-F status
			Signature / Date

Aircraft Daily Log, continued

Flt. No.	Notes (e.g. purpose of flight, weather, sensors used, camera settings, results, lessons learned)

AIRCRAFT DAILY LOG

DATE:

Flt. No.	P-f. I. / C. Cal.	Crew Names		Location		Fuel / Battery			Flight Data						
		Pilot	Other	From	To	Takeoff	Landing	Used	T/o Time	Ld Time	Duration	Ld	Distance	Ht.

Pilot's acceptance prior to today's flight(s): _____

Batteries

Cycles this page:
Cycles brought forward:
Cycles to date:

Flight Data

Totals this page:
Totals brought forward:
Totals to date:

Discrepancies / Incidents

Flt. No.	Description of Technical Issue	Corrective Action Performed (if required)	Approval R-T-F status
			Signature / Date

AIRCRAFT DAILY LOG, CONTINUED

Flt. No.	Notes (e.g. purpose of flight, weather, sensors used, camera settings, results, lessons learned)

AIRCRAFT DAILY LOG

DATE:

Flt. No.	P-f.l. / C. Cal.	Crew Names		Location		Fuel / Battery			T/o Time	Ld Time	Flight Data			
		Pilot	Other	From	To	Takeoff	Landing	Used			Duration	Ld	Distance	Ht.

Pilot's acceptance prior to today's flight(s):

Batteries

Cycles this page:
Cycles brought forward:
Cycles to date:

Flight Data

Totals this page:
Totals brought forward:
Totals to date:

Discrepancies / Incidents

Flt. No.	Description of Technical Issue	Corrective Action Performed (if required)	Approval R-T-F status Signature / Date

Aircraft Daily Log, continued

Flt. No.	Notes (e.g. purpose of flight, weather, sensors used, camera settings, results, lessons learned)

AIRCRAFT DAILY LOG

DATE:

Flt. No.	P-f. I. / C. Cal.	Crew Names		Location		Fuel / Battery			T/o Time	Ld Time	Flight Data			
		Pilot	Other	From	To	Takeoff	Landing	Used			Duration	Ld	Distance	Ht.

Pilot's acceptance prior to today's flight(s): _____

Batteries

Cycles this page:
Cycles brought forward:
Cycles to date:

Flight Data

Totals this page:
Totals brought forward:
Totals to date:

Discrepancies / Incidents

Flt. No.	Description of Technical Issue	Corrective Action Performed (if required)	Approval R-T-F status
			Signature / Date

AIRCRAFT DAILY LOG, CONTINUED

Flt. No.	Notes (e.g. purpose of flight, weather, sensors used, camera settings, results, lessons learned)

AIRCRAFT DAILY LOG

DATE:

Flt. No.	P-f. l. / C. Cal.	Crew Names		Location		Fuel / Battery			T/o Time	Ld Time	Flight Data			
		Pilot	Other	From	To	Takeoff	Landing	Used			Duration	Ld	Distance	Ht.

Pilot's acceptance prior to today's flight(s):

Batteries

Cycles this page:

Cycles brought forward:

Cycles to date:

Flight Data

Totals this page:

Totals brought forward:

Totals to date:

Discrepancies / Incidents

Flt. No.	Description of Technical Issue	Corrective Action Performed (if required)	Approval R-T-F status Signature / Date

AIRCRAFT DAILY LOG, CONTINUED

Flt. No.	Notes (e.g. purpose of flight, weather, sensors used, camera settings, results, lessons learned)

AIRCRAFT DAILY LOG

DATE:

Flt. No.	P-f. I. / C. Cal.	Crew Names		Location		Fuel / Battery			Flight Data					
		Pilot	Other	From	To	Takeoff	Landing	Used	T/o Time	Ld Time	Duration	Ld	Distance	Ht.

Batteries   **Flight Data**

Cycles this page: Totals this page:

Cycles brought forward: Totals brought forward:

Cycles to date: **Totals to date:**

Pilot's acceptance prior to today's flight(s): _____

Discrepancies / Incidents

Flt. No.	Description of Technical Issue	Corrective Action Performed (if required)	Approval R-T-F status
			Signature / Date

AIRCRAFT DAILY LOG, CONTINUED

Flt. No.	Notes (e.g. purpose of flight, weather, sensors used, camera settings, results, lessons learned)

AIRCRAFT DAILY LOG

DATE:

Flt. No.	P-f.I. / C. Cal.	Crew Names		Location		Fuel / Battery			T/o Time	Ld Time	Flight Data			
		Pilot	Other	From	To	Takeoff	Landing	Used			Duration	Ld	Distance	Ht.

Pilot's acceptance prior to today's flight(s):

Batteries

Cycles this page:

Cycles brought forward:

Cycles to date:

Flight Data

Totals this page:

Totals brought forward:

Totals to date:

Discrepancies / Incidents

Flt. No.	Description of Technical Issue	Corrective Action Performed (if required)	Approval R-T-F status
			Signature / Date

AIRCRAFT DAILY LOG, CONTINUED

Flt. No.	Notes (e.g. purpose of flight, weather, sensors used, camera settings, results, lessons learned)

AIRCRAFT DAILY LOG

DATE:

Flt. No.	P-f. I. / C. Cal.	Crew Names		Location		Fuel / Battery			Flight Data					
		Pilot	Other	From	To	Takeoff	Landing	Used	T/o Time	Ld Time	Duration	Ld	Distance	Ht.

Pilot's acceptance prior to today's flight(s):

Batteries		Flight Data	
Cycles this page:		Totals this page:	
Cycles brought forward:		Totals brought forward:	
Cycles to date:		**Totals to date:**	

Discrepancies / Incidents

Flt. No.	Description of Technical Issue	Corrective Action Performed (if required)	Approval R-T-F status Signature / Date

AIRCRAFT DAILY LOG, CONTINUED

Flt. No.	Notes (e.g. purpose of flight, weather, sensors used, camera settings, results, lessons learned)

AIRCRAFT DAILY LOG

DATE:

Flt. No.	P-f. I. / C. Cal.	Crew Names		Location		Fuel / Battery			Flight Data						Approval R-T-F status
		Pilot	Other	From	To	Takeoff	Landing	Used	T/o Time	Ld Time	Duration	Ld	Distance	Ht.	Signature / Date

Pilot's acceptance prior to today's flight(s): _____

Cycles this page: _____
Cycles brought forward: _____
Cycles to date: _____

Batteries **Flight Data**

Totals this page: _____
Totals brought forward: _____
Totals to date: _____

Discrepancies / Incidents

Flt. No.	Description of Technical Issue	Corrective Action Performed (if required)

Aircraft Daily Log, continued

Flt. No.	Notes (e.g. purpose of flight, weather, sensors used, camera settings, results, lessons learned)

AIRCRAFT DAILY LOG

DATE:

Flt. No.	P-f.I. / C. Cal.	Crew Names — Pilot	Crew Names — Other	Location — From	Location — To	Fuel / Battery — Takeoff	Fuel / Battery — Landing	Fuel / Battery — Used	T/o Time	Ld Time	Flight Data — Duration	Flight Data — Ld	Flight Data — Distance	Flight Data — Ht.

Pilot's acceptance prior to today's flight(s):

Batteries — Cycles this page: / Cycles brought forward: / **Cycles to date:**

Flight Data — Totals this page: / Totals brought forward: / **Totals to date:**

Approval R-T-F status

Discrepancies / Incidents

Flt. No.	Description of Technical Issue	Corrective Action Performed (if required)	Signature / Date

Aircraft Daily Log, continued

Flt. No.	Notes (e.g. purpose of flight, weather, sensors used, camera settings, results, lessons learned)

AIRCRAFT DAILY LOG

DATE:

Flt. No.	P.f. I. / C. Cal.	Crew Names			Location		Fuel / Battery					Flight Data				
		Pilot		Other	From	To	Takeoff	Landing	Used	T/o Time	Ld Time	Duration	Ld	Distance	Ht.

Pilot's acceptance prior to today's flight(s):

Batteries

Cycles this page:
Cycles brought forward:
Cycles to date:

Flight Data

Totals this page:
Totals brought forward:
Totals to date:

Discrepancies / Incidents

Flt. No.	Description of Technical Issue	Corrective Action Performed (if required)	Approval R-T-F status
			Signature / Date

AIRCRAFT DAILY LOG, CONTINUED

Flt. No.	Notes (e.g. purpose of flight, weather, sensors used, camera settings, results, lessons learned)

AIRCRAFT DAILY LOG

DATE:

Flt. No.	P.f. I. / C. Cal.	Crew Names		Location		Fuel / Battery			T/o Time	Ld Time	Flight Data			
		Pilot	Other	From	To	Takeoff	Landing	Used			Duration	Ld	Distance	Ht.

Cycles this page:

Cycles brought forward:

Cycles to date:

Batteries

Pilot's acceptance prior to today's flight(s):

Totals this page:

Totals brought forward:

Totals to date:

Flight Data

Discrepancies / Incidents

Flt. No.	Description of Technical Issue	Corrective Action Performed (if required)	Approval R-T-F status
			Signature / Date

Flt. No.	Notes (e.g. purpose of flight, weather, sensors used, camera settings, results, lessons learned)

AIRCRAFT DAILY LOG

DATE: _____

Flt. No.	P-f. I. / C. Cal.	Crew Names		Location		Fuel / Battery			Flight Data					
		Pilot	Other	From	To	Takeoff	Landing	Used	T/o Time	Ld Time	Duration	Ld	Distance	Ht.

Pilot's acceptance prior to today's flight(s):

Batteries

Cycles this page: _____

Cycles brought forward: _____

Cycles to date: _____

Flight Data

Totals this page: _____

Totals brought forward: _____

Totals to date: _____

Discrepancies / Incidents

Flt. No.	Description of Technical Issue	Corrective Action Performed (if required)	Approval R-T-F status	Signature / Date

198

AIRCRAFT DAILY LOG, CONTINUED

Flt. No.	Notes (e.g. purpose of flight, weather, sensors used, camera settings, results, lessons learned)

AIRCRAFT DAILY LOG

DATE: _____

Flt. No.	P-f. I. / C. Cal.	Crew Names		Location		Fuel / Battery			Flight Data					
		Pilot	Other	From	To	Takeoff	Landing	Used	T/o Time	Ld Time	Duration	Ld	Distance	Ht.
														⊠

Pilot's acceptance prior to today's flight(s): _____

Batteries

Cycles this page:	Totals this page:
Cycles brought forward:	Totals brought forward:
Cycles to date:	**Totals to date:**

Flight Data

Discrepancies / Incidents

Flt. No.	Description of Technical Issue	Corrective Action Performed (if required)	Approval R-T-F status Signature / Date

AIRCRAFT DAILY LOG, CONTINUED

Flt. No.	Notes (e.g. purpose of flight, weather, sensors used, camera settings, results, lessons learned)

AIRCRAFT DAILY LOG

DATE: _____

Flt. No.	P-f. I. / C. Cal.	Crew Names		Location		Fuel / Battery			Flight Data						
		Pilot	Other	From	To	Takeoff	Landing	Used	T/o Time	Ld Time	Duration	Ld	Distance	Ht.	

Cycles this page: _____
Cycles brought forward: _____
Cycles to date: _____

Batteries

Pilot's acceptance prior to today's flight(s): _____

Flight Data

Totals this page: _____
Totals brought forward: _____
Totals to date: _____

Discrepancies / Incidents

Flt. No.	Description of Technical Issue	Corrective Action Performed (if required)	Approval R-T-F status Signature / Date

AIRCRAFT DAILY LOG, CONTINUED

Flt. No.	Notes (e.g. purpose of flight, weather, sensors used, camera settings, results, lessons learned)

AIRCRAFT DAILY LOG

DATE:

Flt. No.	P-f.I. / C. Cal.	Crew Names		Location		Fuel / Battery			Flight Data					
		Pilot	Other	From	To	Takeoff	Landing	Used	T/o Time	Ld Time	Duration	Ld	Distance	Ht.

Pilot's acceptance prior to today's flight(s):

Batteries

Cycles this page:

Cycles brought forward:

Cycles to date:

Flight Data

Totals this page:

Totals brought forward:

Totals to date:

Discrepancies / Incidents

Flt. No.	Description of Technical Issue	Corrective Action Performed (if required)	Approval R-T-F status
			Signature / Date

AIRCRAFT DAILY LOG, CONTINUED

Flt. No.	Notes (e.g. purpose of flight, weather, sensors used, camera settings, results, lessons learned)

AIRCRAFT DAILY LOG

DATE: _____

Flt. No.	P-f. I. / C. Cal.	Crew Names		Location		Fuel / Battery			Flight Data					
		Pilot	Other	From	To	Takeoff	Landing	Used	T/o Time	Ld Time	Duration	Ld	Distance	Ht.

Pilot's acceptance prior to today's flight(s):

Batteries

Cycles this page: _____
Cycles brought forward: _____
Cycles to date: _____

Flight Data

Totals this page: _____
Totals brought forward: _____
Totals to date: _____

Discrepancies / Incidents

Flt. No.	Description of Technical Issue	Corrective Action Performed (if required)	Approval R-T-F status Signature / Date

AIRCRAFT DAILY LOG, CONTINUED

Flt. No.	Notes (e.g. purpose of flight, weather, sensors used, camera settings, results, lessons learned)

AIRCRAFT DAILY LOG

DATE: _____

Flt. No.	P-f.I. / C. Cal.	Crew Names		Location		Fuel / Battery			Flight Data					
		Pilot	Other	From	To	Takeoff	Landing	Used	T/o Time	Ld Time	Duration	Ld	Distance	Ht.

Pilot's acceptance prior to today's flight(s): _____

Batteries

Cycles this page: _____
Cycles brought forward: _____
Cycles to date: _____

Flight Data

Totals this page: _____
Totals brought forward: _____
Totals to date: _____

Approval R-T-F status

Signature / Date

Discrepancies / Incidents

Flt. No.	Description of Technical Issue	Corrective Action Performed (if required)

AIRCRAFT DAILY LOG, CONTINUED

Flt. No.	Notes (e.g. purpose of flight, weather, sensors used, camera settings, results, lessons learned)

AIRCRAFT DAILY LOG

DATE: _____

Flt. No.	P-f. I. / C. Cal.	Crew Names		Location		Fuel / Battery					Flight Data				
		Pilot	Other	From	To	Takeoff	Landing	Used	T/o Time	Ld Time	Duration	Ld	Distance	Ht.

Pilot's acceptance prior to today's flight(s):

Batteries

Cycles this page:		Flight Data	Totals this page:
Cycles brought forward:			Totals brought forward:
Cycles to date:			**Totals to date:**

Discrepancies / Incidents

Flt. No.	Description of Technical Issue	Corrective Action Performed (if required)	Approval R-T-F status
			Signature / Date

AIRCRAFT DAILY LOG, CONTINUED

Flt. No.	Notes (e.g. purpose of flight, weather, sensors used, camera settings, results, lessons learned)

AIRCRAFT DAILY LOG

DATE: _____

Flt. No.	P-f. I. / C. Cal.	Crew Names		Location		Fuel / Battery			Flight Data						
		Pilot	Other	From	To	Takeoff	Landing	Used	T/o Time	Ld Time	Duration	Ld	Distance	Ht.	

Pilot's acceptance prior to today's flight(s): _____

Batteries / **Flight Data**

Cycles this page: _____
Cycles brought forward: _____
Cycles to date: _____

Totals this page: _____
Totals brought forward: _____
Totals to date: _____

Discrepancies / Incidents

Flt. No.	Description of Technical Issue	Corrective Action Performed (if required)	Approval R-T-F status Signature / Date

AIRCRAFT DAILY LOG, CONTINUED

Flt. No.	Notes (e.g. purpose of flight, weather, sensors used, camera settings, results, lessons learned)

AIRCRAFT DAILY LOG

DATE: _____

Flt. No.	P-f. I. / C. Cal.	Crew Names		Location		Fuel / Battery			Flight Data					
		Pilot	Other	From	To	Takeoff	Landing	Used	T/o Time	Ld Time	Duration	Ld	Distance	Ht.

Pilot's acceptance prior to today's flight(s):

Batteries

Cycles this page: _____
Cycles brought forward: _____
Cycles to date: _____

Flight Data

Totals this page: _____
Totals brought forward: _____
Totals to date: _____

Discrepancies / Incidents

Flt. No.	Description of Technical Issue	Corrective Action Performed (if required)	Approval R-T-F status
			Signature / Date

AIRCRAFT DAILY LOG, CONTINUED

Flt. No.	Notes (e.g. purpose of flight, weather, sensors used, camera settings, results, lessons learned)

AIRCRAFT DAILY LOG

DATE: _____

Flt. No.	P-f. I. / C. Cal.	Crew Names		Location		Fuel / Battery			Flight Data						Approval R-T-F status
		Pilot	Other	From	To	Takeoff	Landing	Used	T/o Time	Ld Time	Duration	Ld	Distance	Ht.	Signature / Date

Pilot's acceptance prior to today's flight(s): _____

Batteries

Cycles this page: _____
Cycles brought forward: _____
Cycles to date: _____

Flight Data

Totals this page: _____
Totals brought forward: _____
Totals to date: _____

Discrepancies / Incidents

Flt. No.	Description of Technical Issue	Corrective Action Performed (if required)

216

AIRCRAFT DAILY LOG, CONTINUED

Flt. No.	Notes (e.g. purpose of flight, weather, sensors used, camera settings, results, lessons learned)

AIRCRAFT DAILY LOG

DATE:

Flt. No.	P-f. l. / C. Cal.	Crew Names		Location		Fuel / Battery			T/o Time	Ld Time	Flight Data				Approval R-T-F status
		Pilot	Other	From	To	Takeoff	Landing	Used			Duration	Ld	Distance	Ht.	Signature / Date

Pilot's acceptance prior to today's flight(s):

Batteries   Cycles this page:

Cycles brought forward:

Cycles to date:

Flight Data   Totals this page:

Totals brought forward:

Totals to date:

Discrepancies / Incidents

Flt. No.	Description of Technical Issue	Corrective Action Performed (if required)

AIRCRAFT DAILY LOG, CONTINUED

Flt. No.	Notes (e.g. purpose of flight, weather, sensors used, camera settings, results, lessons learned)

AIRCRAFT DAILY LOG

DATE:

Flt. No.	P-f.I. / C. Cal.	Crew Names		Location		Fuel / Battery			T/o Time	Ld Time	Flight Data			
		Pilot	Other	From	To	Takeoff	Landing	Used			Duration	Ld	Distance	Ht.

Pilot's acceptance prior to today's flight(s): _____

Batteries

Cycles this page:
Cycles brought forward:
Cycles to date:

Flight Data

Totals this page:
Totals brought forward:
Totals to date:

Discrepancies / Incidents

Flt. No.	Description of Technical Issue	Corrective Action Performed (if required)	Approval R-T-F status
			Signature / Date

AIRCRAFT DAILY LOG, CONTINUED

Flt. No.	Notes (e.g. purpose of flight, weather, sensors used, camera settings, results, lessons learned)

AIRCRAFT DAILY LOG

DATE: _____

Flt. No.	P-f.I. / C. Cal.	Crew Names		Location		Fuel / Battery			Flight Data						
		Pilot	Other	From	To	Takeoff	Landing	Used	T/o Time	Ld Time	Duration	Ld	Distance	Ht.

Pilot's acceptance prior to today's flight(s): _____

Batteries — Cycles this page: ___ Cycles brought forward: ___ **Cycles to date:** ___

Flight Data — Totals this page: ___ Totals brought forward: ___ **Totals to date:** ___

Discrepancies / Incidents

Flt. No.	Description of Technical Issue	Corrective Action Performed (if required)	Approval R-T-F status — Signature / Date

AIRCRAFT DAILY LOG, CONTINUED

Flt. No.	Notes (e.g. purpose of flight, weather, sensors used, camera settings, results, lessons learned)

AIRCRAFT DAILY LOG

DATE: _____

Flt. No.	P-f. I. / C. Cal.	Crew Names			Location		Fuel / Battery			Flight Data						Approval R-T-F status
		Pilot	Other		From	To	Takeoff	Landing	Used	T/o Time	Ld Time	Duration	Ld	Distance	Ht.	Signature / Date

Batteries

Pilot's acceptance prior to today's flight(s): _____

	Cycles this page:	Totals this page:	
	Cycles brought forward:	Totals brought forward:	
	Cycles to date:	**Totals to date:**	

Flight Data

Discrepancies / Incidents

Flt. No.	Description of Technical Issue	Corrective Action Performed (if required)

AIRCRAFT DAILY LOG, CONTINUED

Flt. No.	Notes (e.g. purpose of flight, weather, sensors used, camera settings, results, lessons learned)

Aircraft Daily Log

Date: _____

Flt. No.	P-f. I. / C. Cal.	Crew Names		Location		Fuel / Battery			Flight Data					
		Pilot	Other	From	To	Takeoff	Landing	Used	T/o Time	Ld Time	Duration	Ld	Distance	Ht.

Pilot's acceptance prior to today's flight(s): _____

Batteries

Cycles this page: _____
Cycles brought forward: _____
Cycles to date: _____

Flight Data

Totals this page: _____
Totals brought forward: _____
Totals to date: _____

Discrepancies / Incidents

Flt. No.	Description of Technical Issue	Corrective Action Performed (if required)	Approval R-T-F status Signature / Date

AIRCRAFT DAILY LOG, CONTINUED

Flt. No.	Notes (e.g. purpose of flight, weather, sensors used, camera settings, results, lessons learned)

Aircraft Daily Log

DATE: _____

Flt. No.	P-f. I. / C. Cal.	Crew Names		Location		Fuel / Battery			Flight Data					
		Pilot	Other	From	To	Takeoff	Landing	Used	T/o Time	Ld Time	Duration	Ld	Distance	Ht.

Pilot's acceptance prior to today's flight(s): _____

Batteries

Cycles this page: _____
Cycles brought forward: _____
Cycles to date: _____

Flight Data

Totals this page: _____
Totals brought forward: _____
Totals to date: _____

Discrepancies / Incidents

Flt. No.	Description of Technical Issue	Corrective Action Performed (if required)	Approval R-T-F status	Signature / Date

Aircraft Daily Log, continued

Flt. No.	Notes (e.g. purpose of flight, weather, sensors used, camera settings, results, lessons learned)

AIRCRAFT DAILY LOG

DATE:

Flt. No.	P-f. I. / C. Cal.	Crew Names		Location		Fuel / Battery			Flight Data						
		Pilot	Other	From	To	Takeoff	Landing	Used	T/o Time	Ld Time	Duration	Ld	Distance	Ht.

Pilot's aceptance prior to today's flight(s): _____

Batteries

Cycles this page:
Cycles brought forward:
Cycles to date:

Flight Data

Totals this page:
Totals brought forward:
Totals to date:

Discrepancies / Incidents

Approval R-T-F status

Flt. No.	Description of Technical Issue	Corrective Action Performed (if required)	Signature / Date

AIRCRAFT DAILY LOG, CONTINUED

Flt. No.	Notes (e.g. purpose of flight, weather, sensors used, camera settings, results, lessons learned)

AIRCRAFT DAILY LOG

DATE:

Flt. No.	P-f. I. / C. Cal.	Crew Names		Location		Fuel / Battery			Flight Data					
		Pilot	Other	From	To	Takeoff	Landing	Used	T/o Time	Ld Time	Duration	Ld	Distance	Ht.

Pilot's acceptance prior to today's flight(s):

Batteries

Cycles this page:

Cycles brought forward:

Cycles to date:

Flight Data

Totals this page:

Totals brought forward:

Totals to date:

Discrepancies / Incidents

Flt. No.	Description of Technical Issue	Corrective Action Performed (if required)	Approval R-T-F status Signature / Date

232

AIRCRAFT DAILY LOG, CONTINUED

Flt. No.	Notes (e.g. purpose of flight, weather, sensors used, camera settings, results, lessons learned)

AIRCRAFT DAILY LOG

DATE:

Flt. No.	P-f. l. / C. Cal.	Crew Names		Location		Fuel / Battery			Flight Data					
		Pilot	Other	From	To	Takeoff	Landing	Used	T/o Time	Ld Time	Duration	Ld	Distance	Ht.

Pilot's acceptance prior to today's flight(s):

Batteries

Cycles this page:

Cycles brought forward:

Cycles to date:

Flight Data

Totals this page:

Totals brought forward:

Totals to date:

Discrepancies / Incidents

Flt. No.	Description of Technical Issue	Corrective Action Performed (if required)	Approval R-T-F status
			Signature / Date

AIRCRAFT DAILY LOG, CONTINUED

Flt. No.	Notes (e.g. purpose of flight, weather, sensors used, camera settings, results, lessons learned)

AIRCRAFT DAILY LOG

DATE:

Flt. No.	P-f. I. / C. Cal.	Crew Names		Location		Fuel / Battery			Flight Data					
		Pilot	Other	From	To	Takeoff	Landing	Used	T/o Time	Ld Time	Duration	Ld	Distance	Ht.

Pilot's acceptance prior to today's flight(s): _____

Batteries

Cycles this page:

Cycles brought forward:

Cycles to date:

Flight Data

Totals this page:

Totals brought forward:

Totals to date:

Discrepancies / Incidents

Flt. No.	Description of Technical Issue	Corrective Action Performed (if required)	Approval R-T-F status
			Signature / Date

AIRCRAFT DAILY LOG, CONTINUED

Flt. No.	Notes (e.g. purpose of flight, weather, sensors used, camera settings, results, lessons learned)

AIRCRAFT DAILY LOG

DATE:

Flt. No.	P-f.I. / C. Cal.	Crew Names		Location		Fuel / Battery			T/o Time	Ld Time	Flight Data			
		Pilot	Other	From	To	Takeoff	Landing	Used			Duration	Ld	Distance	Ht.

Batteries

Pilot's acceptance prior to today's flight(s):

Cycles this page:
Cycles brought forward:
Cycles to date:

Flight Data

Totals this page:
Totals brought forward:
Totals to date:

Discrepancies / Incidents

Flt. No.	Description of Technical Issue	Corrective Action Performed (if required)	Approval R-T-F status
			Signature / Date

AIRCRAFT DAILY LOG, CONTINUED

Flt. No.	Notes (e.g. purpose of flight, weather, sensors used, camera settings, results, lessons learned)

AIRCRAFT DAILY LOG

DATE:

Flt. No.	P-f. I. / C. Cal.	Crew Names		Location		Fuel / Battery			Flight Data					
		Pilot	Other	From	To	Takeoff	Landing	Used	T/o Time	Ld Time	Duration	Ld	Distance	Ht.

Pilot's acceptance prior to today's flight(s): _____

Batteries
Cycles this page:
Cycles brought forward:
Cycles to date:

Flight Data
Totals this page:
Totals brought forward:
Totals to date:

Discrepancies / Incidents

Flt. No.	Description of Technical Issue	Corrective Action Performed (if required)	Approval R-T-F status / Signature / Date

Aircraft Daily Log, continued

Flt. No.	Notes (e.g. purpose of flight, weather, sensors used, camera settings, results, lessons learned)

AIRCRAFT DAILY LOG   DATE:

Flt. No.	P-f. I. / C. Cal.	Crew Names		Location		Fuel / Battery			T/o Time	Ld Time	Flight Data				Approval R-T-F status
		Pilot	Other	From	To	Takeoff	Landing	Used			Duration	Ld	Distance	Ht.	Signature / Date

Pilot's acceptance prior to today's flight(s):

Batteries | Cycles this page: | Flight Data | Totals this page: |
Cycles brought forward: | | Totals brought forward: |
Cycles to date: | | **Totals to date:** |

Discrepancies / Incidents

Flt. No.	Description of Technical Issue	Corrective Action Performed (if required)

242

AIRCRAFT DAILY LOG, CONTINUED

Flt. No.	Notes (e.g. purpose of flight, weather, sensors used, camera settings, results, lessons learned)

AIRCRAFT DAILY LOG

DATE:

Flt. No.	P-f. I. / C. Cal.	Crew Names		Location		Fuel / Battery					Flight Data				
		Pilot	Other	From	To	Takeoff	Landing	Used	T/o Time	Ld Time	Duration	Ld	Distance	Ht.

Batteries / Flight Data

Cycles this page: Totals this page:

Cycles brought forward: Totals brought forward:

Cycles to date: **Totals to date:**

Pilot's acceptance prior to today's flight(s): _____

Discrepancies / Incidents

Flt. No.	Description of Technical Issue	Corrective Action Performed (if required)	Approval R-T-F status
			Signature / Date

Aircraft Daily Log, continued

Flt. No.	Notes (e.g. purpose of flight, weather, sensors used, camera settings, results, lessons learned)

AIRCRAFT DAILY LOG

DATE: _____

Flt. No.	P-f.I. / C. Cal.	Crew Names		Location		Fuel / Battery			T/o Time	Ld Time	Flight Data			
		Pilot	Other	From	To	Takeoff	Landing	Used			Duration	Ld	Distance	Ht.

Pilot's acceptance prior to today's flight(s): _____

Batteries — Cycles this page: _____
Cycles brought forward: _____
Cycles to date: _____

Flight Data — Totals this page: _____
Totals brought forward: _____
Totals to date: _____

Discrepancies / Incidents

Approval R-T-F status

Flt. No.	Description of Technical Issue	Corrective Action Performed (if required)	Signature / Date

AIRCRAFT DAILY LOG, CONTINUED

Flt. No.	Notes (e.g. purpose of flight, weather, sensors used, camera settings, results, lessons learned)

AIRCRAFT DAILY LOG DATE:

Flt. No.	P-f. I. / C. Cal.	Crew Names		Location		Fuel / Battery			Flight Data					
		Pilot	Other	From	To	Takeoff	Landing	Used	T/o Time	Ld Time	Duration	Ld	Distance	Ht.

Pilot's acceptance prior to today's flight(s):

Batteries | Cycles this page: | Cycles brought forward: | **Cycles to date:** _____

Flight Data | Totals this page: | Totals brought forward: | **Totals to date:**

Discrepancies / Incidents

Flt. No.	Description of Technical Issue	Corrective Action Performed (if required)	Approval R-T-F status Signature / Date

AIRCRAFT DAILY LOG, CONTINUED

Flt. No.	Notes (e.g. purpose of flight, weather, sensors used, camera settings, results, lessons learned)

AIRCRAFT DAILY LOG

DATE:

Flt. No.	P-f.l. / C. Cal.	Crew Names		Location		Fuel / Battery			Flight Data					
		Pilot	Other	From	To	Takeoff	Landing	Used	T/o Time	Ld Time	Duration	Ld	Distance	Ht.

Pilot's acceptance prior to today's flight(s):

Batteries
Cycles this page:
Cycles brought forward:
Cycles to date:

Flight Data
Totals this page:
Totals brought forward:
Totals to date:

Discrepancies / Incidents

Flt. No.	Description of Technical Issue	Corrective Action Performed (if required)	Approval R-T-F status Signature / Date

AIRCRAFT DAILY LOG, CONTINUED

Flt. No.	Notes (e.g. purpose of flight, weather, sensors used, camera settings, results, lessons learned)

AIRCRAFT DAILY LOG

DATE:

Flt. No.	P.f.I. / C. Cal.	Crew Names		Location		Fuel / Battery			Flight Data					
		Pilot	Other	From	To	Takeoff	Landing	Used	T/o Time	Ld Time	Duration	Ld	Distance	Ht.

Pilot's acceptance prior to today's flight(s):

Batteries

Cycles this page:
Cycles brought forward:
Cycles to date:

Flight Data

Totals this page:
Totals brought forward:
Totals to date:

Discrepancies / Incidents

Flt. No.	Description of Technical Issue	Corrective Action Performed (if required)	Approval R-T-F status	
			Signature / Date	

252

Aircraft Daily Log, continued

Flt. No.	Notes (e.g. purpose of flight, weather, sensors used, camera settings, results, lessons learned)

AIRCRAFT DAILY LOG

DATE: _____

Flt. No.	P-f. I. / C. Cal.	Crew Names		Location		Fuel / Battery			Flight Data					
		Pilot	Other	From	To	Takeoff	Landing	Used	T/o Time	Ld Time	Duration	Ld	Distance	Ht.

Pilot's acceptance prior to today's flight(s): _____

Batteries

Cycles this page: _____
Cycles brought forward: _____
Cycles to date: _____

Flight Data

Totals this page: _____
Totals brought forward: _____
Totals to date: _____

Discrepancies / Incidents

Flt. No.	Description of Technical Issue	Corrective Action Performed (if required)	Approval R-T-F status
			Signature / Date

254

AIRCRAFT DAILY LOG, CONTINUED

Flt. No.	Notes (e.g. purpose of flight, weather, sensors used, camera settings, results, lessons learned)

AIRCRAFT DAILY LOG

DATE:

Flt. No.	P-f. I. / C. Cal.	Crew Names		Location		Fuel / Battery			Flight Data					
		Pilot	Other	From	To	Takeoff	Landing	Used	T/o Time	Ld Time	Duration	Ld	Distance	Ht.

Batteries
Cycles this page:
Cycles brought forward:
Cycles to date:

Flight Data
Totals this page:
Totals brought forward:
Totals to date:

Pilot's acceptance prior to today's flight(s):
_____

Discrepancies / Incidents

Flt. No.	Description of Technical Issue	Corrective Action Performed (if required)	Approval R-T-F status Signature / Date

Aircraft Daily Log, continued

Flt. No.	Notes (e.g. purpose of flight, weather, sensors used, camera settings, results, lessons learned)

AIRCRAFT DAILY LOG

DATE:

Flt. No.	P-f.I. / C. Cal.	Crew Names		Location		Fuel / Battery			T/o Time	Ld Time	Flight Data			
		Pilot	Other	From	To	Takeoff	Landing	Used			Duration	Ld	Distance	Ht.

Pilot's acceptance prior to today's flight(s):

Batteries

Cycles this page:
Cycles brought forward:
Cycles to date:

Flight Data

Totals this page:
Totals brought forward:
Totals to date:

Discrepancies / Incidents

Flt. No.	Description of Technical Issue	Corrective Action Performed (if required)	Approval R-T-F status
			Signature / Date

Flt. No.	Notes (e.g. purpose of flight, weather, sensors used, camera settings, results, lessons learned)

AIRCRAFT DAILY LOG

DATE: _____

Flt. No.	P-f. I. / C. Cal.	Crew Names		Location		Fuel / Battery			Flight Data					
		Pilot	Other	From	To	Takeoff	Landing	Used	T/o Time	Ld Time	Duration	Ld	Distance	Ht.

Pilot's acceptance prior to today's flight(s): _____

Batteries

Cycles this page: _____
Cycles brought forward: _____
Cycles to date: _____

Flight Data

Totals this page: _____
Totals brought forward: _____
Totals to date: _____

Discrepancies / Incidents

Flt. No.	Description of Technical Issue	Corrective Action Performed (if required)	Approval R-T-F status
			Signature / Date

AIRCRAFT DAILY LOG, CONTINUED

Flt. No.	Notes (e.g. purpose of flight, weather, sensors used, camera settings, results, lessons learned)

AIRCRAFT DAILY LOG

DATE: _____

Flt. No.	P-f. I. / C. Cal.	Crew Names		Location		Fuel / Battery			Flight Data					
		Pilot	Other	From	To	Takeoff	Landing	Used	T/o Time	Ld Time	Duration	Ld	Distance	Ht.

Pilot's acceptance prior to today's flight(s): _____

Batteries — Cycles this page: ___ Cycles brought forward: ___ **Cycles to date:** ___

Flight Data — Totals this page: ___ Totals brought forward: ___ **Totals to date:** ___

Discrepancies / Incidents

Flt. No.	Description of Technical Issue	Corrective Action Performed (if required)	Approval R-T-F status Signature / Date

AIRCRAFT DAILY LOG, CONTINUED

Flt. No.	Notes (e.g. purpose of flight, weather, sensors used, camera settings, results, lessons learned)

Notes Concerning the Aircraft Daily Log

PART 3: Additional Maintenance Actions Log

Additional Maintenance Actions Log

Date	Planned Maintenance / Upgrades / Firmware Updates / Calibrations (other than compass)	Test Date	Approval R-T-F status (sign / date)

ADDITIONAL MAINTENANCE ACTIONS LOG

Date	Planned Maintenance / Upgrades / Firmware Updates / Calibrations (other than compass)	Test Date	Approval R-T-F status (sign / date)

Additional Maintenance Actions Log

Date	Planned Maintenance / Upgrades / Firmware Updates / Calibrations (other than compass)	Test Date	Approval R-T-F status (sign / date)

Additional Maintenance Actions Log

Date	Planned Maintenance / Upgrades / Firmware Updates / Calibrations (other than compass)	Test Date	Approval R-T-F status (sign / date)

Additional Maintenance Actions Log

Date	Planned Maintenance / Upgrades / Firmware Updates / Calibrations (other than compass)	Test Date	Approval R-T-F status (sign / date)

ADDITIONAL MAINTENANCE ACTIONS LOG

Date	Planned Maintenance / Upgrades / Firmware Updates / Calibrations (other than compass)	Test Date	Approval R-T-F status (sign / date)

ADDITIONAL MAINTENANCE ACTIONS LOG

Date	Planned Maintenance / Upgrades / Firmware Updates / Calibrations (other than compass)	Test Date	Approval R-T-F status (sign / date)

Additional Maintenance Actions Log

Date	Planned Maintenance / Upgrades / Firmware Updates / Calibrations (other than compass)	Test Date	Approval R-T-F status (sign / date)

ADDITIONAL MAINTENANCE ACTIONS LOG

Date	Planned Maintenance / Upgrades / Firmware Updates / Calibrations (other than compass)	Test Date	Approval R-T-F status (sign / date)

Notes Concerning the Additional Maintenance Actions Log

Lightning Source UK Ltd.
Milton Keynes UK
UKHW032217260922
409489UK00004B/116

9 782839 925105